KIDS COUNT!

First Steps in Counting

Gary Davis & Catherine Pearn

Republic of Mathematics

Copyright © Gary E. Davis & Catherine A. Pearn

ISBN: 978-0-578-40386-1

First United States edition 2018.

Published by Republic of Mathematics™

Republic of Mathematics™ is an imprint of Krackle LLC

Acknowledgements

We are indebted to Les Steffe for his pioneering research into children's counting, to Bob Wright for his major development and implementation of research in children's counting development in Australia, UK and USA, to Robert Hunting for mentoring both of us into research into young children's mathematical thinking, and to the children who, over many years, have deepened our knowledge and awareness of how children count.

Contents

1 **The Importance of Counting** — 1

2 **How Children Count** — 7
   2.1 Introduction — 7
   2.2 Overview of counting types — 10
      2.2.1 One-to-one correspondence — 10
      2.2.2 Emergent counting — 11
      2.2.3 Physical counting — 13
      2.2.4 Figurative counting — 13
      2.2.5 Counting-on — 13
      2.2.6 Counting by units — 14

3 **Assessment and Assistance** — 15
   3.1 Collecting data — 15
   3.2 Compassionate assessment — 16
   3.3 Informed assistance — 16
   3.4 Keeping a diary — 17

4 **Emergent Counting** — 23
   4.1 Verbal number sequence — 23
   4.2 One-to-one correspondence — 25
   4.3 Observe children carefully — 26

4.4	Assessing for one-to-one correspondence	27
4.5	Emergent Counting	30
4.6	Meaning of numbers for emergent counters	31
4.7	Assessment	32
	4.7.1 How far can they count?	32
	4.7.2 Counting different things	34
	4.7.3 Your assessment	36
	4.7.4 Checklist	37

5 Physical Counting — 39
5.1	Physical Counting	39
5.2	Children, not little adults	40
5.3	Assessment	41
5.4	Meaning of numbers for physical counters	48

6 Assessment is Fuzzy — 51
6.1	Children in transition	51
6.2	Fuzzy assessment	52

7 Learning Through Success — 57
7.1	Learning through success	60
7.2	Behavioral mastery	63
7.3	Focus of attention	64
7.4	Focusing on results and not on actions	64
7.5	Role of the adult	66

8 Helping Emergent Counters — 69
8.1	Nursery rhymes	70
8.2	Counting physical objects	86
8.3	Drawing objects	89
8.4	Recognition of numbers	90
8.5	Dominoes	91

9 Figurative counting — 93
    9.1 What is Figurative Counting? — 95
    9.2 Counting from one — 97
    9.3 Assessment — 98
    9.4 Meaning of numbers for figurative counters — 107

10 Counting-on — 109
    10.1 Counting-on versus counting-all — 109
    10.2 Counting-on & meaning of numbers — 111
    10.3 Parting of the ways — 112
    10.4 You cannot "teach" counting-on — 116

11 Helping physical counters — 121
    11.1 Transition to figurative counting — 121
    11.2 What to look for — 125
    11.3 Further activities — 126

12 Helping figurative counters — 129
    12.1 The burden of counting-all — 129
    12.2 Progressing in figurative counting — 130
    12.3 Moving to counting-on — 133

13 Consolidating counting-on — 141
    13.1 Miscounting by 1 — 141
    13.2 Realizing order does not matter — 141

14 Counting with units — 145
    14.1 Moving from counting-on — 145
        14.1.1 Physical unit counting — 146
        14.1.2 Unit counting in arrays & groups — 148
    14.2 Children's understanding of arrays — 154
    14.3 Separated arrays — 157
        14.3.1 Assessment — 159
        14.3.2 Assistance — 161
    14.4 Figurative unit counting — 164

    14.5 Counting-on by units 167

15 Writing numbers   171
    15.1 Numbers are everywhere 171
    15.2 How children write numbers 174
    15.3 Which is the biggest number? 176
        15.3.1 Numbers in the hundreds 178
    15.4 How many letters are there? 179

16 Addition and subtraction   181
    16.1 Dominoes and number facts 184
        16.1.1 Simple partitions of numbers 184
    16.2 Partitions 185
    16.3 Exploding dots 188
        16.3.1 Base 10 exploding dots : place value 190
        16.3.2 Exploding dots: Addition 193
        16.3.3 Exploding dots: Subtraction 195
        16.3.4 Wait ... there's more! 200

Introduction

This book is based on years of research and experience with many children.

By sharing with you the stories of these many children, we want to help you be awesome: to help your child develop powerful skills in counting, that will serve them well for the rest of their lives.

Understanding and accepting the distilled ideas in this book takes some time and some effort. Is that effort worth it? Will you gain enough benefit that the time spent working through these ideas is worthwhile? We think the answer is:"Yes!"

You've picked up this book, or someone bought it for you, because you, or they, feel a need to help your child with counting. You or they want to help your child to develop strong mathematical skills, to get ahead of the game, to lay strong foundations for their future education.

In this book we place an emphasis on where your child is, and where they could move to from there. We help give you the tools to look carefully at their current counting skills, and figure how to move them to more productive and powerful ways of counting.

This might seem like a simple thing - how hard is counting after all? - but in fact it is not simple. Successful children go through many stages of counting. Many children stay stuck in unproductive strategies. Staying stuck is not helpful as they progress in their education: poor counting skills will hamper them in most subjects and worse, will leave them feeling inadequate.

Let us help you be an awesome parent, one who listens to their kids attempts to count, who knows some things that will help, and who implements helpful means to assist their children develop more productive counting strategies. Happy, confident and skillful children are the result of this awesomeness. It's up to you to be your child's counting hero.

It's our job in writing this book to help you, the heroes, and to give you some extra tools to be as awesome as we know you are.

This book is based on a a network of researchers who have studied how young children develop in their counting skills. It is based on experience with hundreds and thousands of children, teachers and parents who have contributed to this study - who form part of a network of millions upon millions of young children who, daily, learn to count and apply those counting skills in their studies and in everyday life; a network of the millions of parents and teachers who care about these young children and who want the best for them

Gary E. Davis & Catherine A. Pearn, October 2018

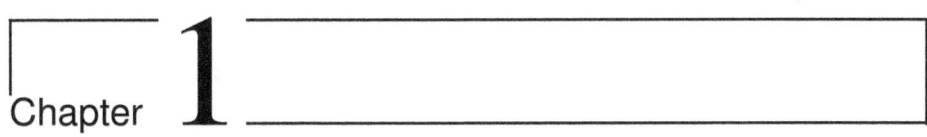

Chapter 1

The Importance of Counting

You want to give your child a head start in mathematics.

You can do this by helping your young child develop deeper counting skills, and deeper understanding of numbers.

At colleges and universities throughout America, and throughout the world, mathematics is the biggest stumbling block for students pursuing higher education, in all disciplines.

If you want your child - your two year old, three year old, four year old, five year old - to succeed in college later you will need to help them lay strong mathematical foundations now. These strong mathematical foundations begin with, and are based in, counting.

We help parents engage deeply with the emotional and intellectual excitement of seeing their children develop efficient numerical skills. The emotional foundations you lay in these early years, through positive feelings of success, will help your child feel confident as they progress through school mathematics and into college.

The ideas we share with are not nearly as widely known as they should be, not even to teachers. Leading educators in the U.S. and Australia have researched these ideas over the many years. Recently they have been tested in schools in the United

States, Australia, and England.

Few elementary teachers, and few college and university faculty who train new teachers, know about the profound impact of this research on young children's counting and mathematical skills.

Teachers who have been exposed to the ideas presented here have responded enthusiastically and have seen the positive impact these ideas have in children's mathematical development.

We work closely with elementary and secondary school teachers. We regularly see students in elementary school, secondary school, and in college, whose counting strategies are inefficient and unsystematic, which bog students down and make mathematics harder than it should be.

Many college students struggle with basic mathematics including algebra.

The groundwork for a deeper understanding of numbers, that underpins all later success in mathematics, lays in efficient counting strategies developed in the early years of life.

Success in obtaining a strong sense of numbers at an early age does not guarantee success at college, but it lays a solid foundation for that later success.

Children who are exposed to a variety of helpful ways of counting in the early years have a better chance of succeeding later at mathematics.

Without some help in counting, children often end up stuck in unhelpful ways of learning mathematics. This can affect their entire school and college development.

Without a firm number sense many older students are doomed to stay at the lower levels of their intended profession.

Daily we see business majors who cannot cope with algebra and calculus at a level required to understand finance.

We see engineering majors who cannot do the algebra required to succeed with calculus.

These sad facts are the basis of much of the failure of young people in college courses - as much as 40% in some engineering courses. Too many students fail in their first choice of professional study because their mathematical skills are woefully deficient.

Teachers at all levels - elementary, secondary, and college - often do not know how to help these students, because they have not been taught the fundamentals of how young children develop numerical skills.

Teenagers and young adults need as much mathematics as they can get to succeed in most professional areas of life. The foundations of their mathematical understanding begin in the earliest years, and it is you - their parents - who can provide the necessary support.

In this book we discuss typical difficulties many children have in counting and arithmetic, and we show how parents can assist children to develop counting skills, and mathematical thinking skills.

Counting is an under-estimated skill. Some people see it as the sort of thing very young children do prior to their moving onto serious topics such as adding, subtracting, multiplying, dividing and more complicated operations with numbers. However counting is a skill that is used in many branches of mathematics, even at an advanced level. It is a skill as important and fundamental as reading, but often not emphasized with the same importance.

Recent research shows that an advanced level of counting is required if children are to be able to learn from current teaching methods in school. But most teachers do not know how to appropriately assess counting skills, and do not know how to tackle deficiencies in counting skills.

You will give your child an advantage if you learn how to assess their counting skills and help them develop more efficient ways of counting. Only then will your child be able to take advantage of current teaching methods common in all schools.

We help you assess your child's current way of counting, and assist you to help them develop the more advanced forms of counting they need to take advantage of school instruction.

Even in the earliest years of school some children find mathematics easy, and some find it hard. The gap between those who find it easy, and those who do not, seems to widen as children get older.

Those who find mathematics hard seem to have a different way of thinking about mathematics that hinders rather than helps them. A deep analysis of what some children do, and others do not, can show us a way forward in helping our children.

We will show you to help your children fulfill their potential with regard to numbers.

We will show you how to observe carefully the ways children count, and how they engage with the adult world of number.

We will show you how to listen to children, and help them progress in their numerical development.

Parents want their children to develop into capable happy adults. Learning to engage with the world of number is part of that development, and begins with children's earliest steps in counting.

We now understand that children progress though clearly defined types of behavior and thinking about counting. These types lead one into the other to a sophisticated ability to count.

Remarkably, recent research shows that many older children, teenagers, and even adults, retain a strong preference for the simpler types of counting behavior that they used in early childhood. These are limited types of counting with which they were comfortable and successful. This restricts their ability to engage more deeply with arithmetic and more advanced mathematics such as high school and college algebra.

So what you do now with your very young child will have a profound effect on their mathematical skills in later years.

We show parents how to help children progress to a more

skilled understanding of numbers.

As children develop in their ability to count, they also develop an understanding of the ways in which we express our number system in speech and writing. Initially, at an early age, this is a parallel development to a child's counting, and comes about mainly because of exposure to written and spoken numbers in a child's environment.

Gradually, around the time a child begins school, these two lines of development come together to form a child's understanding of counting and the system of numbers used by adults.

Not all children develop equally with respect to counting and ways of writing numbers. Many children develop a strong preference for visualizing concrete objects as they count. These children often do not see a need to progress to more efficient ways of counting.

Similarly, many children do not easily see, or grasp, the logical regularities in the numeration system of our culture. For example, many commonly fail to grasp the essence of the base 10 in our numeration system.

Research has shown that practice on procedures alone will not help a child develop an efficient understanding of number. To the contrary, evidence suggests that an exclusive focus on taught procedures and routines for counting, for arithmetic, and for understanding place value, lead only to failure. This is the way to produce only inflexible robots, and not supple, agile minds.

Children begin by doing things in their own way. They learn the adult culture's ways by imitation and by use in context. As in other aspects of human life, so it is in counting and arithmetic.

We will assist you to look at, and listen to, children as they take their early steps in number.

We will help you see clearly the development that is taking place, where that development fits into a larger scheme of things, and what you can do to help your child develop strong and useful numerical abilities.

Counting is an excellent example of the use of cultural tools to help children develop higher levels of awareness.

The number words "one", "two", "three", "four" ... (in English) are more or less arbitrary cultural signs that a child learns to use as they carry out counting actions. These number-words come to have deeper meaning for children as they use them to count things more efficiently and productively.

Your role as parent is critical in their development as capable, mathematically proficient, young people.

Chapter 2

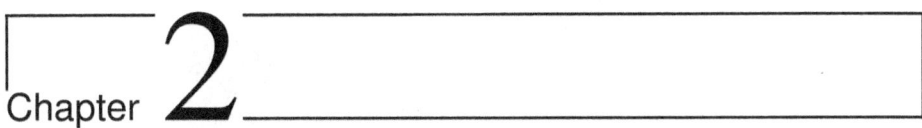

How Children Count

2.1 Introduction

What do children do when they count? What sort of actions do they carry out? What do they think about when they are counting?

In this book - Kids Count: First Steps in Counting - we base our answers to these questions mainly on the pioneering work of Leslie P. Steffe (pronounced "Steffy"), at the University of Georgia, and his former student Robert Wright, at Southern Cross University in Australia.

Steffe has made a deep study of the types of children's counting behavior. His research is based on ideas of the Swiss psychologist Jean Piaget (pronounced "Zhohn Pea-AH-zhay"; see figure 2.2, below). Steffe has worked with numerous colleagues in his quest to unravel children's mathematical thinking and development.

Leslie P. Steffe   Robert J. Wright

Figure 2.1: Les Steffe and Bob Wright

Figure 2.2: Jean Piaget

As a result of this work we can now distinguish a number of different counting types - different ways in which children count. Robert Wright and colleagues, working out of Southern Cross University in northern New South Wales, Australia, have developed, and applied, these counting types in classrooms in the United States, Australia and England. Steffe and his co-workers developed the basis for the counting types through keen observation of children. They used Jean Piaget's fundamental ideas of children's mental development to help them see the counting types clearly.

The counting types have proved to be remarkably stable. The counting types predict children's counting development in a wide variety of settings, and in every country in which they have been tested.

The counting types will be explained below. We will describe typical behavior of children in each of the counting types.

We will show you how to find out what type of counter your child is right now.

We give you ideas on how to help your child strengthen their counting strategies.

Then we give you some important ideas on how you can help your child progress to more efficient and productive counting strategies.

2.2 Overview of counting types

The counting types develop by Leslie Steffe and co-workers describe behaviors of children as they develop deeper understanding of counting and of numbers.

In this book we look at the counting types up to "counting-on", in which a child can count on from a given number, such as "7, 8, 9, 10" to figure that 3 more than 7 is 10.

The reason that counting-on is so important is that it signals a major step in a child's counting development. The ability to count-on and to use counting-on in different settings is what points to a child beginning to develop a strong and flexible sense of numbers.

2.2.1 One-to-one correspondence

Learning to count effectively is not simple for very young children. To be successful with counting they need to co-ordinate several actions. For example, to count a group of objects children need to know the following:

1. The number names, in correct order, up to the number of objects in the group.

2. That the last number they say is the actual number of the objects in the group.

3. That they can start counting from any object in the group.

4. That the objects in the group need not be identical.

5. That they count each object in the group once and only once.

We refer to an ability to count each object in a group once and only once as being able to make a one-to-one correspondence. This means that a child can co-ordinate each number name with precisely one object in the group of things they are counting.

2.2.2 Emergent counting

Children three, four, and even five, years old will sometimes not yet be able to say the number names in correct order. Often they will have difficulty matching number names with objects.

We call these children *emergent counters*.

An emergent counting will show beginning signs of being able to count - to match number names with objects, using one-to-one correspondence. But they will not be consistent in their counting. An emergent counter might successfully count 5 objects today, yet tomorrow they might not be able to count 4 objects correctly.

We use the term "emergent counter" in this book as an abbreviation for "a child who uses emergent counting".

It is important to keep in mind that the term "emergent counter" is not a characteristic description of a child, and should not be used as a label for a child.

It is simply an assessment of a child's predominant counting actions and abilities at a particular time in their life.

A child who uses emergent counting strategies will, sometime soon, develop into physical counting, then figurative counting, and counting-on.

Their counting *development* is key, as is recognizing where they are in that development.

2.2.3 Physical counting

Children who use physical counting can do things that those who use emergent counting cannot.

They can count small numbers of physical objects around the house.

A child who uses physical counting may be able to consistently count up to 5 objects but may not, yet, be able to count 25 objects.

What is important is whether they can consistently count up to a certain number of objects - 5 for example.

A physical counter will show consistency in their counting. Typically there will be counting tasks, such as counting as much as 45 objects, that are simply too hard, yet, for a beginning physical counter.

2.2.4 Figurative counting

There are things that a physical counter cannot do. A physical counter cannot count hidden objects.

For example, if there are 3 cookies on the table in plain view, and 2 more in the pantry, out of sight, a physical counter cannot figure out how many there are in total.

A child who can consistently work out how many objects there are in total when some of those objects are hidden from sight is called a figurative counter.

Figurative counters will always count from 1 when they need to find the total number of objects, such as 5 cookies and 3 more that are hidden from sight.

2.2.5 Counting-on

When a child finds the total of 5 cookies and 3 more by counting "six, seven, eight" they are counting-on.

The counting-on strategy signals a major re-organization of a child's thinking about counting.

Children who use counting-on are able to recall that a number word like "five" is not only the number after four in the number word sequence - it is also a signal for memories of having counted "one, two, three, four, five".

So, for children who can count-on, the number words have a double meaning - both as a word in the number word sequence and, importantly, as a trigger for recall of previous acts of counting.

It is those explicit memories that allow them to begin counting from six, for example. A figurative counter, by contrast, would have to begin counting again at one.

2.2.6 Counting by units

Children develop a great flexibility in their counting strategies when they can count unit bundles. Often this begins by counting in 2's: "2,4,6,8, ..." and then in 5's: "5, 10, 15, 20, ..."

Counting unit bundles relies on a child seeing a collection of things - for example two apples - as a unit bundle, as a single thing to be counted, and not as two separate apples to be counted individually.

Initially, children will count unit bundles physically, often tapping them as they count, just as they would with individual items. Then they progress to being able to count hidden unit bundles, and then to counting-on with unit bundles.

Chapter 3

Assessment and Assistance

This book has a very important theme: learning to collect data through compassionate assessment, for informed assistance.

3.1 Collecting data

We could have subtitled this book "Evidence Based Parenting". That is because a major feature of this book, one that runs throughout every chapter, is learning how to collect evidence on your child's counting skills, and how to use that evidence to help your child progress to more advanced counting skills.

Without evidence the greatest danger is that you will rush your child into trying to do something that is presently too hard for them.

With a careful assessment of what they can do now you can help them do better in the future.

That assessment requires paying careful attention to the evidence for their present counting skills.

You have to learn what to look for, and that is a major part of what we focus on in this book.

Then you have to take that evidence of their counting sills

and figure how to help your child progress. In doing that, the evidence you have collected on their present counting skills is critical. With that evidence you can help your child to always feel successful and confident in their counting development.

3.2 Compassionate assessment

Compassionate assessment means, first and foremost, assessment of your child's counting strategies right here and now.

We call this compassionate assessment because it is assessment done with love and care: assessment that aims not to simply judge, but to provide help, in the most informed way possible.

In this book you will find structured sequences of tasks that help you assess your child's counting skills.

These tasks fit into a model of counting types. The counting types are distilled from an analysis of many children's counting activities. They provide a framework in which children's development in counting can be seen more clearly. Use them, as we explain, and you will see with clarity how your child is progressing in counting.

3.3 Informed assistance

Informed assistance means help that you can give your child that is based on accurate assessment of what they can do now.

We give many examples of informed assistance based on years of experience with young children as they progress in counting.

A most important aspect of informed assistance is to give children counting tasks that are a quantum out of their reach - just enough to stimulate them to think, bu t no so much as to overwhelm them.

In order to know what sort of counting tasks to give your child to help them develop their counting skills, you need to have an accurate picture of what they can do at any given time.

As your child counts they will develop deeper and deeper meanings for numbers. They will do this with your informed help.

In the beginning their meaning for numbers is based on saying number names and touching objects to be counted.

As they progress, with your help, they will acquire much deeper understanding of numbers through successively more efficient ways of counting.

To give them this help you need to learn how to compassionately assess their counting at any given moment, and then how to help them move forward based on your assessment.

3.4 Keeping a diary

We will show you how to assess where your child is in their development of counting skills.

When you assess where they are, you will want to follow up with suggested strategies to help reinforce and strengthen their skills. You will then want to help move them to new and more productive counting strategies.

As you assess and then help your child develop new skills you will find it very helpful to keep a diary of their growth and development.

In our experience it is very easy to forget exactly what a child could do last week. Keeping a diary is a big help in making accurate assessments. These assessments form the basis of your further plans to help your child develop new counting strategies.

A diary can simply be an exercise book with enough room for you to make brief notes, or draw simple pictures of what your child did on a given day.

Figure 3.1: A useful counting diary

Simply make a brief note, with the date, that you can refer to later.

For example you might write in the diary as follows:

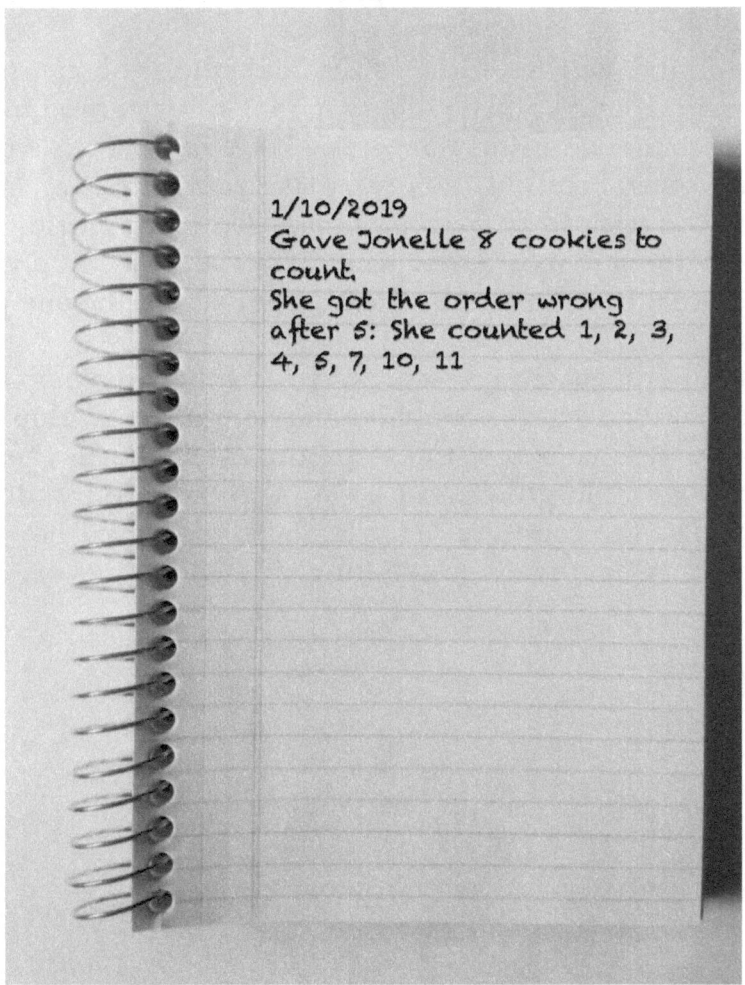

Figure 3.2: Enter the date & activity in the diary

This will help you tremendously when you want to see what effect your assessments and help are having for your child.

The diary will be a record of their progress as you help them develop counting skills. Remember, the diary entries need only be brief.

If you and Jonelle did more that day, then you can make a record of it. If that's all you did, then that's all you need to write.

We recommend using an exercise book rather than a regular diary for several reasons. You can make your entries as small or large as you wish in an exercise book. You make entries as you have something to record, not necessarily every day. And you can make several brief entries for successive days on the one page if you wish.

The counting diary is a record of your work with your child. It records how you assess your child and help them develop counting skills. It is your tool to use productively and to cherish as you see your child grow and develop their counting skills. You can also use the counting diary to show your child how much progress they have made in counting.

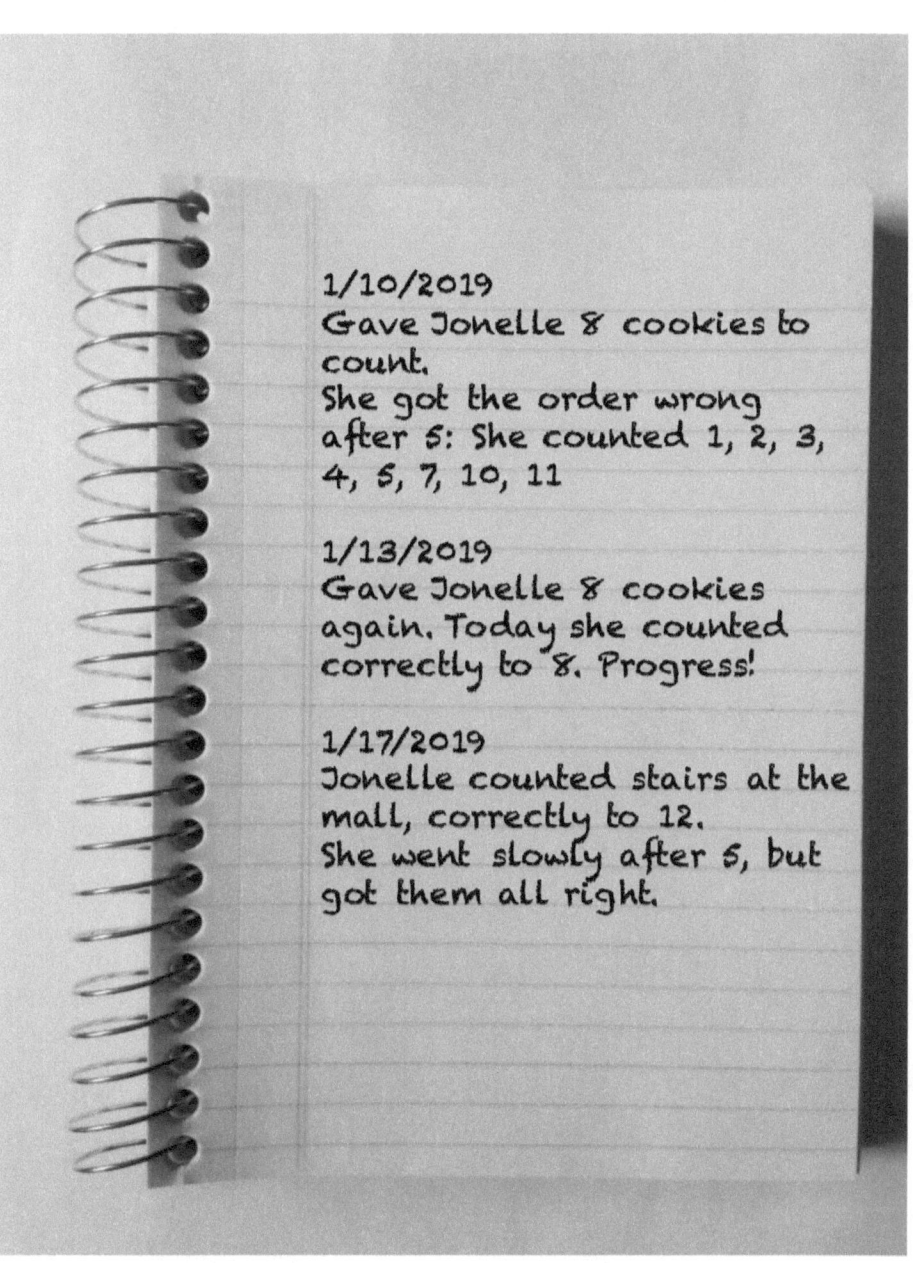

Figure 3.3: Enter your observations in the diary

Chapter 4

Emergent Counting

4.1 Verbal number sequence

Sally is at the beach with her mother, Marion. Sally is 3 years old and she and her mother are going to make sand-castles. As they walk down the stairs to the beach, Marion asks Sally if she can count the stairs. Carefully, Sally says the number names "one", "two", three", … as she puts each foot on a new stair.

Saying the number names in order - "one", "two", "three", "four", "five", … - is the most important first step for a child to learn to count. In the beginning a child's success is dependent on their experience in saying number names in order. Their ability to remember the number names in order is very much like learning the alphabet.

Figure 4.1: Sally counting steps

Just as Sally and her mother counted steps, there are many everyday opportunities for children to practice saying the number names in order.

In our experience children will happily show you how good they are at counting. For example children can, and will, practice counting at breakfast, in the car, on the way to school, coming home, at meals and snack times, and in the bath at night. Just a couple of minutes of your time each day listening to them count will help your child become more skilled at saying the number names in order.

The number names are what a child builds upon as they progress from physical counting of things they can see, hear, or touch, to more powerful forms of counting.

All children first need to count physical objects: things that can be touched, seen or heard, that can be picked up and moved, or repeated, such as clapping sounds.

Physical objects include things around the house, for example, apples, toys and cookies. Physical objects also include things outside the house for example, leaves, rocks and shells. At school children will use physical objects such as counters, blocks, and popsicle sticks.

For your child to be a successful counter of physical objects they need to be good at reciting the number names in order.

They also need to use one-to-one correspondence, which we now discuss in more detail.

4.2 One-to-one correspondence

The basis of counting is that we match up one collection of things with another: cookies with children, for example.

Figure 4.2: Matching cookies and kids

But when children match cookies one-to-one with people neither the cookies nor the people are usually arranged in straight lines. A child therefore has to track who has received a cookie:

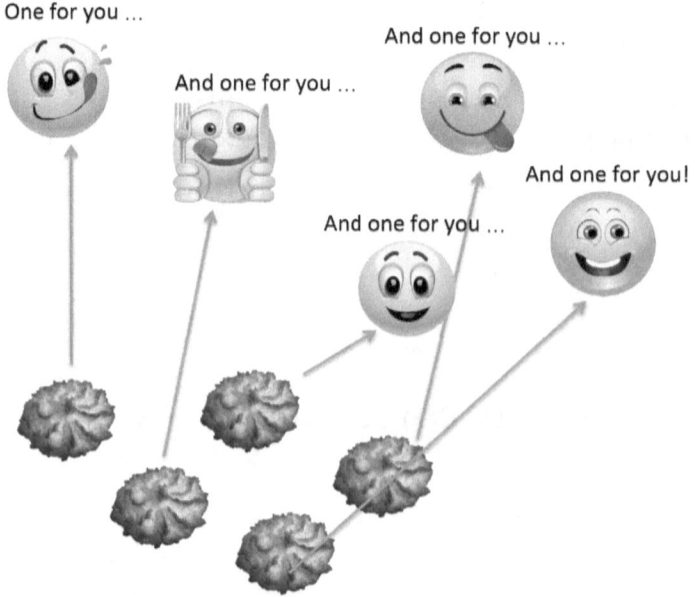

Figure 4.3: Tracking who gets a cookie

Pay careful attention to how your child counts: how they track, what they touch, what they say.

Record as much as you can remember in their diary. The diary entries help you develop too: they help you begin to see more clearly what it is that your child is doing as they count. This knowledge is essential as you help them develop more efficient types of counting.

4.3 Observe children carefully

Sometimes children will say numbers in the correct order, as if they were counting, but in fact they are not establishing one-to-one correspondence as you might think.

For example, Adalira is walking up the library steps with her sister, and she is saying "one, two, three, " as she walks up the steps.

But her counting does not match the steps - sometimes she counts twice on one step, and sometimes does not say a number word at all as she makes a step.

Adalira is not establishing a one-to-one correspondence between number names and steps.

When 3 year old Sally is counting steps, she sometimes steps over two stairs, but only does one count: "One" (one stair), "two" (one stair), "three" (one stair), "four" (two stairs), "five" (one stair).

What is Sally doing?

She is matching the number words with her actions of stepping. She is not, yet, matching the number words with the physical stairs.

Sally has a form of one-to-one correspondence because she matches the number words correctly with her stepping actions. But she is not, yet, matching the number words with the stairs.

4.4 Assessing for one-to-one correspondence

Very young children can show some success in counting, but easily lose their way because they are still learning how to count physical objects.

Their ability to establish one-to-one correspondence can be easily stressed with larger numbers.

You can very simply test your child's grasp of one-to-one correspondence.

For example, Tonya placed 8 cookies in front of her daughter, Isabella.

Tonya asked Isabella to tell her how many cookies were on the table. Tonya watched very carefully to see if Isabella touched or pointed to each cookie while saying a number name.

Tonya was not sure what Isabella was doing, so she said: "Isabella, show me how you counted those cookies. Tell me out loud

Figure 4.4: Eight cookies, placed higgledy piggledy

what you are doing."

Isabella had difficulty counting 8 cookies. She could count 3 okay, but got lost saying the number words and touching cookies after that.

Tonya took away the cookies and put just 3 in front of Isabella. She then asked Isabella to tell her how many cookies there were now on the table.

Isabella moved a cookie and said "one". She moved another cookie and said"two" and then moved the remaining cookie and said"three".

Isabella could use one-to-one correspondence for the 3 cookies but she had difficulty with one-to-one correspondence for larger numbers of cookies. Now that Tonya knows Isabella can use one-to-one correspondence for 3 cookies, but not for 8 cookies, she will try Isabella with 4 and then 5 cookies.

Isabella's sister Marietta, who is one year older than Isabella, could easily count 8 cookies using one-to-one correspondence.

She pointed one by one at the cookies and said, "one, two, three, four, five, six, seven, eight" as she pointed at each cookie.

Tonya placed 12 cookies on the table for Marietta to count, and watched carefully to see whether she still used one-to-one correspondence to count the larger number of cookies. She did, so Tonya tried 15 cookies for Marietta to count.

Counting activities like those used by Tonya with her two children allow you to see whether your child is able to use one-to-one correspondence for larger numbers of objects.

At first your child will likely move each individual object, as Isabella did, and later touch or point to them, as Marietta did. They will do this to keep track that each object is given only one number name, as they count.

As children become more expert at using one-to-one correspondence it will get harder for you to tell what strategy they are using to keep track of objects.

Sometimes your child will nod their head as they track the objects to be counted.

They might nod their head, blink, or move their eyes as they track the objects.

It is often difficult to see what they are doing. Their movements can be very slight.

Just watch and observe carefully the movements your child makes as they count. These movements, often quite small, are a key to your child's further development in counting.

Be sure to record these movements in your child's counting diary.

Some children who are saying number names correctly are not yet counting. What they are doing is preparing to count. They are learning the number names with the help of a parent and how to match those names with objects in the world.

We call this activity emergent counting, and children who do it consistently, emergent counters.

4.5 Emergent Counting

Tom is three years old and says he can count. He proudly tells any one who will listen: "I can count. 1, 2, 3, 5, 4, 10."

His older brother Sam is four and is also very proud of his counting. "I can count too: 1, 2, 3, 4, 5, 6, 7, 8, 9, 10, 20, 30, 40, 50, 60, 70, 80, 90, 20."

Both boys have shown some success with saying the number names in order, but they will need more practice saying them in correct order.

Their mom has placed 8 bananas on the table in front of them.

Figure 4.5: Counting bananas, arranged higgledy piggledy

When she asked the boys how many bananas are on the table, Tom's answer was not correct, because he had not yet learned to say the number names in the correct order.

Sam was also incorrect because he was not yet able to match the number names, in order, with the bananas on the table.

The boys are using emergent counting. Emergent counting means that children try the counting sequence but are not yet able to count physical objects.

Tom does not realize that saying the number names in order helps him to count objects. He is concentrating on saying the number names correctly and getting them in correct order.

Sam is not yet skilled enough to say the number names in correct order while touching or pointing at the objects one at a time.

4.6 Meaning of numbers for emergent counters

The meaning of numbers for emergent counters comes from their memories of reciting the number names, sometimes in correct order, sometimes not.

Emergent counters will also have memories of adults or older children, asking them to say the number names and match the number names with objects, such as crackers.

The meaning of numbers is very limited for emergent counters - it is based on their recollections of actions of saying number names and moving, pointing at, or focusing on, objects while they try to say the number names, one at a time. Sometimes they do it in correct order, sometimes not.

Sofia's mother asked if Sofia could count the stairs as they walked from one level of the mall to the next.

"Yes! I can do it by myself!" she said.

As she lifted her feet and placed them on the stairs she counted: "One, two, three, five, seven, six!"

Sofia could not yet establish a one-to-one correspondence between the number names and stairs, and she could not say the number numbers up to six in correct order.

Sofia's understanding of numbers is based on trying to say the number names and trying to match that with her actions. That is how she remembers numbers and counting, and that is what they mean to her at this point in her development of counting.

Unlike reciting the alphabet, counting involves saying the number names in correct order as well as matching those number names with actions such as stepping on stairs, or moving or touching objects.

Sofia recalls that is what she should do as she counts. Her understanding of numbers is limited to trying to say the number names correctly and match them with actions.

4.7 Assessment

4.7.1 How far can they count?

Peter said to his son Glenn: "Can you count by ones for me as far as you can?"

Glenn did not seem sure what his dad meant, so Peter said: "Like this: 1, 2, 3..." and waited for Glenn to do it after him.

Figure 4.6: Father checking how far child can count

Peter listened to find the highest number to which Glenn could recite the number names correctly. Glenn said: "one, two three, four, five, six, seven, eight, eleven, nine, ten!"

The number 8 was as far as Glenn could count correctly by ones.

Peter recorded this number in Glenn's counting diary, along with the date, and a note that he asked Glenn to count as high as he could. This record will help Peter check if Glenn always does the same thing or if he does something different next time.

Next, Peter wanted to see if Glenn could count things around the house.

He started with 4 cookies, placing them on the table in front of Glenn and asking: "How many cookies are on the table?"

Glenn gave an instant answer of 4 and appeared to have guessed. Peter wasn't sure that Glenn had counted, so he asked: "Can you count them?"

Sometimes your child will give an instant response but you are not sure how they worked it out. If that happens, you can simply ask: "How did you do that?" Children are usually happy to tell an adult how they counted.

Peter watched carefully to see what Glenn did. He was looking to see if Glenn:

- Touched or moved each cookie as he counted it.

- Counted each cookie only once.

- Said the counting sequence correctly.

Peter again made a brief written note in Glenn's counting diary, because he knows this will help him see more clearly how Glenn's counting strategies change over time.

Glenn was successful counting 4 cookies. But what if he had not been successful? What could Peter do then? A sensible strategy is to try again with a smaller numbers of cookies, for example 2 or 3.

Glenn was successful counting 4 cookies so Peter tried again with larger numbers of cookies - 5, 6, and 7 - which Glenn could successfully count. Peter recorded this in Glenn's counting diary.

4.7.2 Counting different things

Monica wanted to see if her daughter Haley could count different sorts of things, so she got a set of farm animals that included sheep, cows and pigs.

Monica was thinking she could also use different fruit such as apples, oranges, pears, and bananas.

She placed 10 of the toy animals in front of Haley, and asked: "Haley, how many animals are there on the table?"

Figure 4.7: 10 toy farm animals

Haley did not an answer right away, and seemed to be thinking, so Monica asked her: "Can you count them?"

Haley then said: "Ten!"

Because Monica was not sure how Haley got the answer of ten she asked: "How did you do that?"

Monica then watched very carefully to see what Haley did. She was looking to see if Haley:

- Touched or moved each animal as she counted it.

- Counted each animal only once.

- Used the counting sequence correctly.

If your child is unsuccessful counting 10 animals try again with smaller numbers of animals, for example, 3, 4 or 5.

If they are successful at counting these smaller numbers of animals then try again with slightly larger numbers, say 12, 13 or 14.

4.7.3 Your assessment

When you have given these tests for emergent counting to your child you will have discovered one of four things:

1. Your child was unable to state the conventional counting sequence and did not use one-to-one correspondence when counting the objects.

2. Your child used the correct counting sequence but they were unsuccessful, because they did not use one-to-one correspondence to determine the number of objects.

3. Your child used one-to-one correspondence when counting the objects, but their counting sequence was not the conventional one. That is, they left a number out or they said the numbers in the wrong order.

4. Your child can consistently count a small number of objects successfully using one-to-one correspondence. This means that each object was counted exactly once. The counting sequence was correct. In this case you child is not using emergent counting. They have progressed beyond emergent counting to a more efficient form of counting.

If your child is not consistently successful at counting a small number of objects then they are probably still using emergent counting. This might be because their counting sequence is not correct. Or it might be because they are unable to consistently use one-to-one correspondence when counting physical objects.

4.7.4 Checklist

1. How far can your child say the number names in correct order?

2. How many identical objects on a table can your child count?

3. How many non-identical objects on a table can your child count?

4. How many identical objects can your child count out - for example, spoons the child places on a table - without your help?

5. How many non-identical objects - for example, knives, forks, spoons - can your child count out without your help?

6. When your child counts do they move, touch, or point to objects? If not, do they nod or blink as they keep track of objects? What body movements do they make as they keep track of objects?

7. When counting, can your child start from any object on a table? Can they start from any object and count correctly when the objects are in a straight line, or in a circle, or placed randomly?

Chapter 5

Physical Counting

5.1 Physical Counting

Jonelle is four years old, and she can count. She can say the number words in correct order up to 20, and she can count up to 9 things accurately. Beyond that she sometimes makes mistakes in matching the number names with the things she is counting.

Jonelle says the teen numbers very carefully because she is not quite sure how they go, and she doesn't yet know what comes after 29.

Jonelle can count cookies and other things around her home if she can see and touch them. But if her mother gives her a plate of cookies and tells Jonelle there are three more in the pantry, Jonelle cannot figure how many there are in all. She needs to see or touch cookies to count them.

Right now, Jonelle uses only physical counting, so we call her a *physical counter*.

This means that right now she needs to see, feel, or hear physical objects in order to count them.

Jonelle will develop her counting skills so that one day - maybe tomorrow, maybe next week - she will be able to count hidden objects.

For now, Jonelle is assessed as a physical counter, but soon she will develop other counting skills.

A child who is a physical counter is limited to counting those things they can feel, see or hear.

A physical counter will point to and touch cookies as they count them. They will point to fish in a tank to count them; and they will listen to clapping sounds to count how often a parent claps.

Children who are physical counters must feel, see, or hear the things they count. Like Jonelle, they cannot count objects that are hidden from their senses.

A lot of five, six and seven year old children are physical counters.

5.2 Children, not little adults

A child takes a big step forward in their counting when they can count things that they cannot, touch or hear.

Adults are often surprised when children cannot count hidden objects. Most adults no longer need to see and touch small collections of objects, for example up to 20, in order to count them.

As adults, we know that if there are 8 cookies on a plate and 3 more in the pantry then there are 11 in all.

At some point in our past we learned how to count hidden objects. It is difficult for us to understand how a young child cannot see what is so obvious to us.

Trying to understand children's thinking, by observing and interpreting what they do, has been one of the important developments in mathematical education.

When we listen to children and understand their ways of thinking, we greatly increase our ability to help them grow and to be successful.

5.3 Assessment

Alina Spinillo, of the Federal University of Pernambuco, Brazil, working with Peter Bryant, at Oxford University in England, devised a simple and revealing test to identify physical counters.

Figure 5.1: Alina Spinillo and Peter Bryant

This test alone will not tell you with certainty that a child can only count physical objects, but it is a strong indicator. Used with other similar tests, described below, it can convince us that a child needs to feel, see, or hear, physical objects in order to count them.

Here is our version of the test. To begin, you will need:

- 18 crackers (or other food substitute, such as jelly-babies).

- 3 small dolls (you can get them at many department stores or toy stores).

- 1 cardboard box (a shoe box will do) with a slit in one side, large enough to place one of the crackers, but not so large as to allow a child to easily see through.

- A large sheet of paper or card to cover some of the crackers so the child cannot see them and cannot feel them.

Figure 5.2: Eighteen crackers, to be shared evenly between three dolls

Then proceed as follows:

> Put the 3 small dolls on a table or on the floor.
>
> Say to the child: "The dolls are very hungry. They want to eat all the crackers. But they all want the same - a fair share. Can you fix it so they all get the same?"
>
> As the child picks up the crackers, and just before they share them out, pick up one of the dolls and say: "Wait - this doll is going off to her room."
>
> Place the doll in the box so that the doll is out of view of the child.
>
> Say: "Can you fix it so that the doll in here gets her fair share too?"
>
> If the child shares out some, but not all of the crackers, say that the dolls are very hungry and want all the crackers.
>
> When the crackers are all shared out, completely cover the crackers of one doll with the piece of paper, or card, and ask the child to count the crackers of the other doll outside the box.
>
> Now ask: "How many crackers does the doll in the box have?"

Now here is a most important point: this is a diagnostic test. As you give this test to your child you are NOT trying to help them get it right.

Uh-oh! Does this sound wrong, or what? Aren't you, as parent, supposed to be helping your kid learn how to do things right?

Yes, you are. But there's a bigger picture.

To help them do things right you first need to know accurately how they do things now. And then you need to think very carefully about that before doing anything.

Your role, right now, is to compassionately assess what your child does. In order to best help your child, you need to know accurately what they can do by themselves: you need data!

Your job right now, at this moment, is NOT to help them succeed in figuring out how many crackers the doll in the box has: that's their job. Remember - you already know how many crackers the doll has in the box. The question you want answered is: Does my child know this?

If they do not, then no amount of teaching, telling, or showing them, at this stage will help.

We will show in Chapter 11 how to help a physical counter learn to count hidden objects. But if a child is rushed they will only fail, or lose interest because they see an adult can do what they cannot.

Spend a great deal of care and attention in simply watching, listening, and learning to assess where they are, right now. Then, with the aid of this book, help them move on.

As you carefully watch your child, something miraculous will happen. Your child will come to believe, because it is true, that you really care about how they think, that you are paying attention to THEM.

Because of your attention, in which you are listening and watching, but not telling them what to do, they will be much more willing to share their thoughts with you, now and in the future. You will become more deeply invested in their growth, and they will appreciate you as a caring, listening parent.

When Karen did this test with her son Michael she found that he was very confident the doll in the box had 6 crackers. Karen asked him how he knew, and Michael answered with certainty: "Because they all have the same!"

If this is what your child says, then the Spinillo-Bryant test is not conclusive, and your child may have progressed to a more sophisticated type of counting. More tests are needed to decide this.

Joe Becker, at the University of Illinois at Chicago, devised a similar test. In this variant of the test, a child again has to share crackers to three dolls, but this time all dolls are visible.

To begin this task, you will need:

- 15 crackers (or other food substitute).

- 3 small dolls.

- Two large sheets of paper or card.

You proceed as follows:

> Put the 3 small dolls on a table or on the floor.
>
> As before, say to your child: "The dolls are very hungry. They want to eat all the crackers. But they all want the same - a fair share. Can you fix it so they all get the same?"
>
> If your child shares out some, but not all of the crackers, say that the dolls are very hungry and want all the crackers.
>
> When the crackers are all distributed, completely cover the crackers of two of the dolls with the pieces of paper, or card.
>
> Point to the doll whose crackers are uncovered, and ask: "How many crackers does this doll have?"
>
> When the child has counted, or answered without counting, ask: "How many crackers do the other two dolls have?"

Physical counters cannot answer this last question.

Do not try to help your child answer as you assess where they are - just watch what they do. At this point you are building up observational skills in order to better help them. Helping comes a little later.

Another variant of the hidden objects counting task was devised by Robert Hunting at East Carolina University, and his former student Kristine Pepper, then at La Trobe University in Australia.

Figure 5.3: Robert Hunting

They asked children to equally share coins - one lot to a place where they could see the coins, the other into a piggy bank with a slot through which the coins can be placed, but not seen. As with the other two tests, you can ask the child how many coins they can see, and then how many are in the piggy bank. A physical counter cannot answer this question because for them counting means seeing and touching or pointing at the coins as they say the number names.

A single use of any one of these tests will not conclusively show a child to be a physical counter. However, the use of several tests will be revealing. If a child cannot answer the "How many?" question for hidden objects on several tests, then you should have a very strong suspicion that the child can only count physical objects. The likelihood that they can count hidden objects is very low.

Often children will appear to be able to count objects they cannot see when in fact they are seeing them. In Becker's test, for example, unless the hidden objects are truly hidden, children can peep under the card or cloth without an adult knowing. Sometimes we have been surprised that a child cannot answer "How many?" in the Spinillo-Bryant test, but can in Becker's test. When we ask: "How did you do that?" children will tell us that they were peeping under the card. So we need to be careful - children want to answer the "How many?" question and they will try all they know or can do to answer it correctly.

Physical counters face a big problem in imagining how to count things they cannot see. These tests are designed to find out if, at a particular time and aspect of development, that is where a child is in their counting.

When you have given these tests for physical counting to your child you will have discovered one of three things:

- Your child can count hidden objects without fail.

- Your child can sometimes count hidden objects, but they are often unsure and their success is patchy.

- Your child cannot count hidden objects at all.

In the first case you child has progressed beyond physical counting: you should no longer think of them as a physical counter. What this means is that they are consistently able to count things they cannot see, hear, or feel. They have progressed beyond physical counting to a more efficient form of counting called figurative counting.

5.4 Meaning of numbers for physical counters

The meaning of numbers for physical counters lies in the memory of the last number they say as they count physical objects.

Having moved from emergent counting to stable physical counting, children are now able to reliably make one-to-one correspondences to answer the question: "How many?" Their answer is shown in the last number name they say.

Annie, a five-year old, was writing how many people would fit in the seats of a drawing of a cinema. She had reached 27 when an older friend, Jamie, asked her: "Which number are you up to, Annie?"

Annie answered: "Twenty seven."

Jamie then asked: "How many people would fit in those seats you've written on?"

Annie counted, from the beginning: "One, two, three,..., twenty-six, twenty-seven."

"Twenty-seven", she said.

Annie could answer the "How many?" question by counting from one, using one-to-one correspondence, but she was not aware that the number "27", which she herself had just written, was also the answer to the "How many?" question. Annie's meaning for numbers did not yet extend to knowing that a number name is the number of objects there are in a collection.

This might seem very odd, because didn't Annie just count "one, two,... twenty-seven"? and then say: "Twenty-seven" to answer the question: "How many people could fit on those seats?" Yes, she did. But Annie knows to do this only in the context of actually counting "one, two, three,...".

When she is asked a question such as Jamie asked, where the "27" Annie wrote is not in a direct counting context, Annie does not know that the number 27 she has just written is the answer to the question of how many people are seated. To her, the question: "How many?" is a trigger to count, starting from one. The question "How many?" means for Annie to begin counting from 1.

Chapter 6

Assessment is Fuzzy

6.1 Children in transition

This chapter contains a "health warning". The message is this: do not expect assessment of your child to be clean cut and simple, based on one or two assessment tasks.

Any description of child development that tries to place children at various "levels" of development suffers from the obvious dilemma that children will sometimes be in transition from one level to the next. Their behavior at these times shows characteristics of both levels, so making it hard to place them accurately. What's more, children can show behaviors characteristic of three, or more, different levels at one and the same time. This is especially so when they are given counting tasks that are subtly different, and challenge their current abilities in different ways.

Although it is convenient, we would prefer not to label children as "emergent counters" or "physical counters". Giving them labels like this makes it sound as though the behavior the child shows is a characteristic of the child. We know that children develop, and show other, more sophisticated, counting behaviors. And children can show behaviors from different counting types at the same time.

6.2 Fuzzy assessment

The four-year old girl, Till, whose counting behavior we look at below shows some characteristics of an emergent counter, a lot of characteristics of a physical counter, and hints of characteristics of a figurative counter - a child who can count hidden objects. So which is she? Is she a physical counter in transition to being figurative counter? If so, why does she show some behavior typical of an emergent counter?

On balance, we think she is best described as a physical counter, with some behaviors indicating a movement to figurative counting, and other behaviors still indicative of emergent counting. What this tells us is that she needs the basic skills of one-to-one correspondence, used in emergent counting, worked on and strengthened. She also needs more practice doing physical counting.

The example is taken from the research of our colleague Robert Hunting, then at East Carolina University.

In what follows detail an account of Robert's interviews with Till.

This information is provided courtesy of Robert Hunting and PALM-Ed: Partnership for Advancing the Learning of Mathematics http://www.palm-ed.com.

We recommend you visit the PALM-Ed site to learn more about young children's mathematical development.

Figure 6.1: Till begins counting bugs

Robert was talking with Till, who was, at that time, 4 years and 11 months old. On the table in front of them were 4 large plastic bugs. Robert first asked Till how many bugs there were on the table: "How many bugs have we got today?" he asked.

Till picked up the bugs, moved them from her right side to her left side, counting "one, two, three, four" as she moved the bugs one at a time.

So far so good: as a result of her counting, Till knew there were 4 bugs on the table.

Robert then reminded Till that they had been talking about how it had started to rain and that one little bug wanted to hide under a red rock.

There was a small red cup placed upside down on the table that they pretended was a red rock.

Till placed one of the bugs under the rock. Robert asked Till how many bugs she could see now. Till tapped one of the bugs and said "one". She tapped the next bug twice, saying "two, three" as she did so, and then tapped the third bug, saying "four."

Robert asked again "How many?" Till touched a bug and said "one, two", touched the next bug and said "three", and then touched the last bug and said "four."

Are we now sure that Till's counting answered the question "How many?" for her?

We might think that Till was remembering her previous count to 4 when there were 4 bugs on the table, and that's the answer she expected if she did the counting actions over.

If this reasoning is correct then her memory of counting would be much stronger than her actions to count, because both times there were 3 bugs she said two number words when touching only one bug. It's as if she knew what the answer should be so she had to very quickly adjust her counting to make sure she got that answer.

Robert asked Till how many bugs were hiding under the rock, and she held up a finger and said "one."

Robert now said another little bug wanted to hide under the rock. Till pushed a bug under the rock and Robert asked: "So now, how many little bugs can you see?"

Till held up two fingers and then touched the two bugs on the table one at a time, saying "one, two" as she did. Robert asked how many bugs were hiding under the red rock and Till held up two fingers, saying "two".

Robert said another bug wanted to get dry, and Till placed another bug under the rock. Robert asked how many bugs she could see now. Till had her hand on the bug and was moving it around the table. She hesitated for a moment, took her hand off the bug, then pushed it and said "Only one."

Robert asked Till how many bugs were under the rock. She held up several fingers and said "one, two, three."

What was Till counting? Did she remember how many bugs she placed under the rock and was mentally counting them off, as if they were images in her head? Till is showing signs of moving beyond physical counting because she can tell us how many bugs were under the rock, without being able to see them. But before we get too excited at Till's ability to count thing she cannot see, let's continue with the story.

Till put the last bug under the rock and Robert asked her how many bugs were out in the rain now. Till replied: "None."

Robert then asked her how many bugs were hiding under the rock. Till responded: "I don't know"

This is a response typical of a physical counter. But we must be careful judging Till to be a physical counter because she has shown some evidence of being able to count things she cannot see. In some respects Till seems to be a counter who is in transition from being a physical counter to the next counting type of figurative counting. Yet, as we have seen, Till miscounted 3 bugs as 4 - a behavior typical of an emergent counter

The moral of this story of Till's counting is to watch children carefully as they count. Pay attention to what they are doing, learn to record their actions, and do not rush to judgment.

Equally, when you are watching them to assess where they are, do not rush to help them. As tempting as it is, your helping them as you test them only interferes with your ability to obtain a reliable estimate of their abilities. You want to help your child grow and develop. So spend time with these simple fun tests to figure out first where they are in counting. Then tomorrow, or the day after, you can help them develop new counting skills.

Chapter 7

Learning Through Success

How do children learn new counting skills? How do they move from one set of counting behaviors to another?

The counting types give us a solid framework on which to assess children's counting abilities. But how do we help them develop more efficient ways of counting?

Our answer to these questions is based on the work of the British developmental psychologist Annette Karmiloff-Smith.

Karmiloff-Smith worked with Jean Piaget, whose theories of development form the basis for Steffe's counting types.

Central to Piaget's ideas on the development of children's thinking is interaction with other people.

These interactions can lead a child to come into conflict with their own ways of thinking or doing things, and stimulate the child to accommodate other ways of acting and thinking.

Figure 7.1: Annette Karmiloff-Smith

Piaget's ideas of conflicts in thinking can be seen clearly in Steffe's work on children's counting.

For example, when we present a physical counter with a task such as figuring the total of 5 cookies they can see on the table and 4 more in the pantry that they cannot see, we set up a potential disturbance in the child's thinking.

If the child cannot yet do this task, they are put into a state of conflict: their wish to work out the total of the cookies is frustrated by the fact that they still have to see the cookies in order to answer the question: "How many?"

The idea of conflicts in children's thinking forms the basis of <u>assessment</u> of children's counting. We give children carefully designed counting tasks to see what they can and cannot do at any given time. We carefully observe what a child can do, and what frustrates them.

Karmiloff-Smith developed the idea that children can learn under conditions of success, without necessarily experiencing conflict, or the need to resolve those conflicts.

Her ideas form the basis of <u>assisting</u> children to develop more efficient ways of counting.

After careful assessment, our focus is on how to move children on to more efficient forms of counting.

Following Annette Karmiloff-Smith's lead, we do this when they are already showing strong signs of success in their current counting skills.

Piaget's ideas of conflict are useful in assessing children.

Karmiloff-Smith's ideas of learning under conditions of success seem to be more productive in helping children develop new skills.

Think of it this way:

1. Setting up a potential conflict for a child allows us to assess their counting skills.

2. Building on a child's successful counting skills, in new situations, allows us to help them develop more efficient skills.

7.1 Learning through success

Rachel was playing a counting game with her eight year old son Logan and three of his friends - Andrew, Abigail and Paige. Rachel had placed a numbered hundreds board on the table.

She asked the children to find where 37 would be on the board.

They all counted by ones from 1.

Rachel asked them to find 10 more than 37, and again they all counted by ones from 1.

1	2	3	4	5	6	7	8	9	10
11	12	13	14	15	16	17	18	19	20
21	22	23	24	25	26	27	28	29	30
31	32	33	34	35	36	37	38	39	40
41	42	43	44	45	46	47	48	49	50
51	52	53	54	55	56	57	58	59	60
61	62	63	64	65	66	67	68	69	70
71	72	73	74	75	76	77	78	79	80
81	82	83	84	85	86	87	88	89	90
91	92	93	94	95	96	97	98	99	100

Figure 7.2: Hundreds board showing counting by 1's to 37 & then 10 more

She said: "So ten more than 37 is 47."
She then asked them to find 10 more than 47.
Again they all counted by ones from 1.
Rachel then said: "37 and 10 more is 47, 47 and 10 more is 57, so what would 10 more than 57 be?"
Abigail tapped twice, moved her finger down to the row below 57, on the 67 spot, and grinned.
The other three children counted by ones.
Rachel then said: "37 and 10 more is 47, 47 is 10 more than 47 is 57, and 10 more than 57 is 67, so what would 10 more than 67 be?"
Now Abigail just moved her finger straight down to the next row.
Logan began to count by ones, and then moved his finger down.
Andrew and Paige counted by ones.
Rachel asked Abigail how she got the answer.
Abigail said: "It follows a pattern. 37, 47, 57, 67, 77."
Rachel then asked: "What is 10 more than 77 ?"
Abigail and Logan immediately moved their fingers to 87 in the row below 77, and Andrew and Paige continued to count by ones.
Rachel said: "37 and 10 more is 47, 10 more than 47 is 57, 10 more than 57 is 67, 10 more than 67 is 77, 10 more than 77 is 87, so what would 10 more than 87 be?"
The children did the same as for 10 more than 77.
Finally, Rachel asked them: "What would 10 more than 97 be?"
Abigail and Logan both answered: "107". Andrew and Paige said they did not know.
In this example we see the heart of the idea of learning through success as it applies to counting:

- The activity allowed all four children to use their mastery of physical counting to find the numbers Rachel asked them, on the hundreds board.

- Abigail and Logan's mastery of physical counting freed up their brains enough so they could change their focus of attention.

Instead of focusing on carrying out physical counting, which they could do very well, their minds now began to focus on what they were doing as they counted.

This shift of attention made it clear to Abigail, and then to Logan, that they were just moving down the hundreds board to the next row.

How simple! But this simple observation required their minds to be free enough to pay attention to it.

They could change their focus of attention because physical counting was so easy for them.

Abigail's tapping and pointing is how she explicitly described, to Rachel and to herself, what she was doing as a result of counting. It is also how Rachel could see that Abigail had changed her focus of attention.

Abigail and Logan represented "10 more than" in a new way. At first they responded to this task by counting by ones, from 1. Through being able to re-focus on what they were doing as a result of that counting by ones, they could find a new interpretation of "10 more than" - it now meant moving down to the next row in the hundreds board.

Abigail and Logan's success was very specific to this problem, and to the use of the hundreds board. We have just a little evidence that Abigail and Logan have begun to be able to count in units of 10 because they could answer that 10 more than 97 is 107, without seeing 107 on the hundreds board, and without any physical counting.

Abigail and Logan did not learn a general way of counting by tens and then apply it to the hundreds board: they began to learn the idea of counting by tens from thinking about their own physical counting in the very specific context of the hundreds board. The observations they made for the specific counting task, in a specific context, led to a more general way of thinking, not the other way around.

7.2 Behavioral mastery

A central feature of learning through success is *behavioral mastery*. Children are much more likely to show signs of development under conditions of success when they are skilled in carrying out actions in a prior stage of their development.

For example, children who are emergent counters, and not yet consistently able to count physical objects, are much more likely to become physical counters if their present skills are strengthened. These skills include such things as reciting the number names correctly.

The reason for this is simple and has to do with a function of our brains called working memory. Working memory is that aspect of our brains that deals with holding and using information in real time. When a child can carry out an activity without much thought, their working memory is freed to think about what they are doing as they do it. This is known as mental reflection, and it is a key idea underlying Steffe's counting types.

This means that practice at perfecting skills is very important. But mere practice alone is not enough. The practice must be thoughtful and appropriate. It must be done for a good reason. That way, practice becomes a powerful tool to free a child's thinking processes.

7.3 Focus of attention

What moves children to a new stage of development in their counting is their own thinking, based on what they remember.

Especially important is how they think about what they have been doing in their previous counting activities.

This inner reflection signals a change in the child's focus of attention.

We can see this clearly even at the earliest steps of learning to count.

Emergent counters are focused on trying to say the number names in correct order.

They are focused on their own actions, and how well they are learning the number names and number sequence.

When they can do this well there is now a possibility that they can begin to focus on saying those number names as they touch objects one at a time.

This happens as their focus of attention shifts from their efforts to say the number names correctly to saying the number names and touching objects.

What allows this possibility is their behavioral mastery of saying the number names.

7.4 Focusing on results and not on actions

When you are helping your child to count there is a most important distinction you should make. When you give your child counting tasks they will carry out those tasks according to their skill level. However, at each level, what your child is not yet able to do is to take the result of that counting as a thing to think about.

For example, Sarah - who was a more advanced counter - could count quickly by twos. Her mother asked her to count a bunch of chips on the table and Sarah quickly said: "two, four, six, eight, ten, twelve, fourteen, sixteen, eighteen."

Her mother asked: "Sarah, how may times did you count by two?"

Sarah could not answer that question when her mother asked it. She was simply unaware of how many times she counted by twos. She was only aware that she had counted by twos to get an answer of eighteen.

Her focus of attention was on her actions of counting by twos to get an answer to the question: "How many chips?"

When her mother asked her how many times she counted by twos she could not say, because that was not the focus of her attention.

We may wonder why it is that a child who is good at counting by twos, for example, cannot, or does not, take the result of that counting as something to think about.

The reason seems to be that the child's working memory is occupied with the act of counting, and does not have space to think about those acts.

It is only when the acts of counting become routine for your child that their working memory is sufficiently free to allow them a chance to think about the result of their counting.

For Sarah to tell how many times she counted by twos she will have to have behavioral mastery of counting by twos and be able to re-focus her attention from counting by twos to counting how many times she counts by twos.

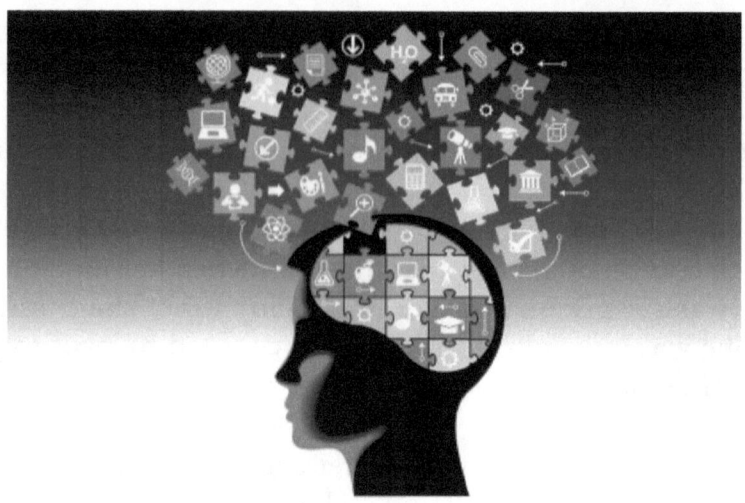

Figure 7.3: Schematic illustrating working memory fully occupied

7.5 Role of the adult

A parent has an important role to play in stimulating a child's focus of attention in counting.

A child might change their focus of attention themselves. However, counting is a cultural activity. We are not born knowing how to count, and we learn to do it through interaction with people who are experienced and skilled at counting, usually parents. A parent has a big role to play in stimulating a child to reflect on their actions.

In the example, above, Rachel plays an important role in focusing the thinking of the four children Logan, Andrew, Abigail and Paige, on the actions of counting they have just carried out. By summarizing what they have done, Rachel gently flags them to pay attention to the outcome of their counting actions.

Parental involvement cannot and should not be rushed. It is simply not possible to "teach" a child to count more efficiently if they have not yet obtained behavioral mastery of their current counting actions.

For example, Katrina's grandson Jack tried to count MMs with about 14 or 15 in a box.

Jack could successfully count up to 10 MMs but then waved his hands and said 11, 12, 13, without establishing one-to-one correspondence.

His mom watched and said: "That's wrong, Jack. Count them again."

Jack did, and got a different answer, because he was not establishing one-to-one correspondence after 10 MMs.

"Count them, Jack, count them", said his mom in frustration, trying to get Jack to touch the MMs as he said the number words.

But Jack could not because at age 3 years 3 months he could only establish one-to-one correspondence up to 10 objects.

Kristina advised her daughter to be patient, and spend time watching what Jack does, rather than rushing him to do something he cannot do properly yet.

Jack needs more work on consolidating his skills of saying the number names in correct order. When he can do that, and when saying the bigger numbers like 11, 12 and 13 are easier for him, Jack will not have to pay much attention to how he says the numbers. Then, and only then, will his brain have space in working memory for him to pay attention to establishing one-to-one correspondence with MMs and number names.

Chapter 8

Helping Emergent Counters

There are three main reasons why emergent counters have difficulties counting groups of objects.

1. One reason is the difficulties they have with the verbal counting sequence. Your child might leave numbers out while they count, for example, when counting four objects they may say: 1, 2, 4, 5.

2. Another reason is that your child might say the numbers in the wrong order. For example, when counting five objects they may say: 1, 2, 4, 3, 5. In this example the child gets the right answer and if you had not been listening carefully you would not be aware that they have problems counting. However, if you had given them 3 or 4 objects to count they might have got the wrong answer.

3. Another difficulty for emergent counters is that they are often unable to consistently use one-to-one correspondence when counting physical objects. They will sometimes get the right answer for small groups of objects but it depends on the number of objects and how fast or slow they say the counting sequence.

Emergent counters can usually recite some number names. Their success is dependent on their ability to remember a series of number names. Often they do not realize that counting is done for the purpose of determining how many things there are in all.

Emergent counters need to strengthen their ability to count out loud and to count larger numbers of physical objects before they can develop new counting proficiency and become physical counters.

8.1 Nursery rhymes

Emergent counters can, and will, develop physical counting. To help your child you should reinforce their current skills by practicing the verbal counting sequence.

A good place to start is with the numbers from 1 through 5. Once your child is successful to 5, increase the numbers from 1 through 10.

Nursery rhymes and singing rhymes are always useful, especially when they are done with appropriate actions. There are many examples of nursery rhymes and singing rhymes loved by young children and their parents.

We have reproduced some examples of well-known nursery rhymes on the following pages.

The BBC has approximately 20 videos of counting songs to which you can sing along: https://bbc.in/2Mm5qq0

Super Simple Learning has a YouTube video of 10 popular counting songs: https://bit.ly/2xzdloc

One, Two, Buckle My Shoe:

1, 2, buckle my shoe

One, two,
Buckle my shoe;
Three, four,
Shut the door;
Five, six,
Pick up sticks;
Seven, eight,
Lay them straight:
Nine, ten,
A big, fat hen.

Five Little Ducks:

Five little ducks went out one day

Five little ducks
Went out one day
Over the hill and far away
Mother duck said
"Quack, quack, quack, quack."
But only four little ducks came back.

Four little ducks
Went out one day
Over the hill and far away
Mother duck said
"Quack, quack, quack, quack."
But only three little ducks came back.

Three little ducks
Went out one day
Over the hill and far away
Mother duck said
"Quack, quack, quack, quack."
But only two little ducks came back.

Two little ducks
Went out one day
Over the hill and far away
Mother duck said
"Quack, quack, quack, quack."
But only one little duck came back.

One little duck
Went out one day
Over the hill and far away
Mother duck said
"Quack, quack, quack, quack."
But none of the five little ducks came back.

Sad mother duck
Went out one day
Over the hill and far away
The sad mother duck said
"Quack, quack, quack."
And all five little ducks came back.

Baa Baa Black Sheep:

Baa, baa, black sheep

Baa, baa, black sheep,
Have you any wool?
Yes, sir, yes, sir,
Three bags full;
One for the master,
And one for the dame,
And one for the little boy
Who lives down the lane.

One, Two, Three, Four, Five:

One, two, three, four, five
Once I caught a fish alive

One, two, three, four, five,
Once I caught a fish alive,
Six, seven, eight, nine, ten,
Then I let it go again.
Why did you let it go?
Because it bit my finger so.
Which finger did it bite?
This little finger on the right.

Five little speckled frogs:

Five little speckled frogs

Five little speckled frogs
Sat on a speckled log
Eating the most delicious bugs
Yum yum
One jumped into the pool
Where it was nice and cool
Then there were four green speckled frogs.

Four little speckled frogs
Sat on a speckled log
Eating the most delicious bugs
Yum yum
One jumped into the pool
Where it was nice and cool

Then there were three green speckled frogs.

Three little speckled frogs
Sat on a speckled log
Eating the most delicious bugs
Yum yum
One jumped into the pool
Where it was nice and cool
Then there were two green speckled frogs.

Two little speckled frogs
Sat on a speckled log
Eating the most delicious bugs
Yum yum
One jumped into the pool
Where it was nice and cool
Then there was one green speckled frog.

One little speckled frog
Sat on a speckled log
Eating the most delicious bugs
Yum yum
The frog jumped into the pool
Where it was nice and cool
Then there were *no* green speckled frogs.

Ten green bottles hanging on the wall:

Ten green bottles hanging on the wall

Ten green bottles hanging on the wall
Ten green bottles hanging on the wall
And if one green bottle should accidentally fall,
There'll be nine green bottles hanging on the wall.

Nine green bottles hanging on the wall
Nine green bottles hanging on the wall
And if one green bottle should accidentally fall,
There'll be eight green bottles hanging on the wall.

Eight green bottles hanging on the wall
Eight green bottles hanging on the wall
And if one green bottle should accidentally fall,
There'll be seven green bottles hanging on the wall.

Seven green bottles hanging on the wall

Seven green bottles hanging on the wall
And if one green bottle should accidentally fall,
There'll be six green bottles hanging on the wall.

Six green bottles hanging on the wall
six green bottles hanging on the wall
And if one green bottle should accidentally fall,
There'll be nine five bottles hanging on the wall.

Five green bottles hanging on the wall
Five green bottles hanging on the wall
And if one green bottle should accidentally fall,
There'll be four green bottles hanging on the wall.

Four green bottles hanging on the wall
Four green bottles hanging on the wall
And if one green bottle should accidentally fall,
There'll be three green bottles hanging on the wall.

Three green bottles hanging on the wall
Three green bottles hanging on the wall
And if one green bottle should accidentally fall,
There'll be two bottles hanging on the wall.

Two green bottles hanging on the wall
Two green bottles hanging on the wall
And if one green bottle should accidentally fall,
There'll be one green bottle hanging on the wall.

One green bottle hanging on the wall
One green bottle hanging on the wall
And if one green bottle should accidentally fall,
There'll be *no* green bottles hanging on the wall.

Five fat sausages:

Five fat sausages sizzling in a pan

Five fat sausages sizzling in a pan
The grease got hot - and one went "BANG"!

Four fat sausages sizzling in a pan
The grease got hot - and one went "BANG"!

Three fat sausages sizzling in a pan
The grease got hot - and one went "BANG"!

Two fat sausages sizzling in a pan
The grease got hot - and one went "BANG"!

One fat sausage sizzling in a pan
The grease got hot - and it went "BANG"!

No fat sausages frying in a pan!

Five little monkeys:

Five little monkeys

Five little monkeys jumping on the bed
One fell off and bumped his head
Mama called the doctor and the doctor said:
"No more monkeys jumping on the bed!"

Four little monkeys jumping on the bed
One fell off and bumped his head
Mama called the doctor and the doctor said:
"No more monkeys jumping on the bed!"

Three little monkeys jumping on the bed
One fell off and bumped his head
Mama called the doctor and the doctor said:
"No more monkeys jumping on the bed!"

Two little monkeys jumping on the bed
One fell off and bumped his head

Mama called the doctor and the doctor said:
"No more monkeys jumping on the bed!"

One little monkey jumping on the bed
He fell off and bumped his head
Mama called the doctor and the doctor said:
"No more monkeys jumping on the bed!"

Now there's no little monkeys jumping on the bed.
They're all jumping on the sofa instead!

Ten in a bed:

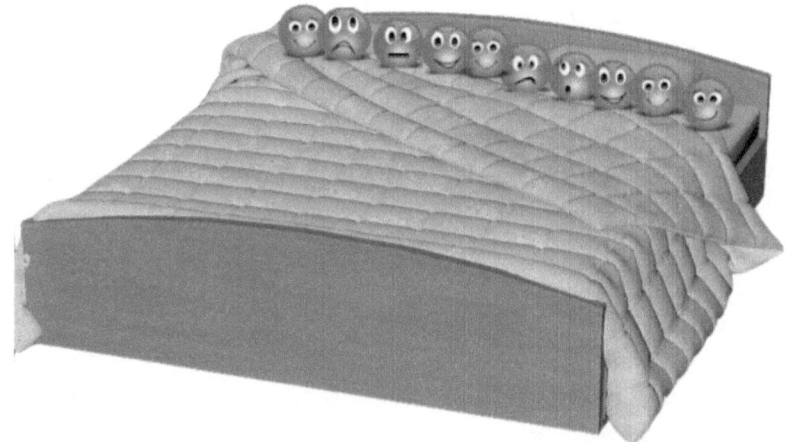

There were ten in a bed

There were ten in a bed
And the little one said
"Roll over, roll over"
So they all rolled over
And one fell out

There were nine in a bed
And the little one said
"Roll over, roll over"
So they all rolled over
And one fell out

There were eight in a bed
And the little one said
"Roll over, roll over"
So they all rolled over
And one fell out

There were seven in a bed
And the little one said
"Roll over, roll over"
So they all rolled over
And one fell out

There were six in a bed
And the little one said
"Roll over, roll over"
So they all rolled over
And one fell out

There were five in a bed
And the little one said
"Roll over, roll over"
So they all rolled over
And one fell out

There were four in a bed
And the little one said
"Roll over, roll over"
So they all rolled over
And one fell out

There were three in a bed
And the little one said
"Roll over, roll over"
So they all rolled over
And one fell out

There were two in a bed
And the little one said
"Roll over, roll over"
So they all rolled over

And one fell out

There was one in a bed
    And the little one said
    "Good night!"

8.2 Counting physical objects

As you practice the verbal counting sequence, encourage your child to count physical objects.

You can start by asking your child to count parts of their bodies starting with a small numbers of things.

For example, ask how many hands (feet, ears, eyes, legs, fingers on one hand, toes on one foot) they have.

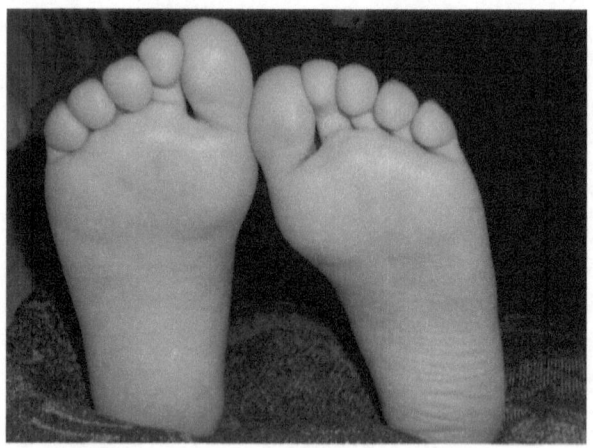

Figure 8.1: How many toes?

Ask your child to point to the parts as they count, because emergent counters need practice matching things they are counting with the number names.

You can ask your child to count objects around the house.

For example: How many apples are on the table? How many cups are on the sink? How many spoons are on the table for dinner? The possibilities are limited only by your imagination! You will have a lot of fun doing counting like this.

Figure 8.2: Count the apples

Watch to see how they check they have the right number of objects. If you cannot tell by watching, just ask your child: "How do you know you have the right number?" Typically they will count again to be satisfied they got it right.

Once your child is comfortable counting small numbers of objects you need to ask your child to show you a certain number of things.

For example, ask your child to place 6 spoons on the table. Ask them to show you things such as 4 apples, 5 crayons, or 2 bath toys. Again, watch to see how they check that they have the right number of objects, and if you cannot tell by watching then ask, for example: "How do you know you have 6 spoons?"

For example, Stephanie asked her son Andrew to place 4 spoons on the table. Stephanie watched Andrew concentrate as he placed spoons on the table. She could not tell by watching how he knew he had 4 spoons, so she asked him: "Andrew, how do you know you have 4 spoons?"

Andrew then counted; "One, two, three, four. See, four!"

Stephanie knew that Andrew could count to 4, but she still was not sure how he knew he had 4 spoons as he was placing them on the table. So Stephanie asked Andrew if he could place 6 forks on the table. "Can you tell me out loud what you are doing?" she asked.

Figure 8.3: Count the forks

Andrew picked up the bundle of forks and placed them one at a time on the table, in a setting. As he did so, he counted out loud: "One, two, three, four, five, six. See, six!"

Stephanie was happy that she now knew how Andrew counted

out the forks.

8.3 Drawing objects

Once your child can show you a number of physical objects such as 4 apples, 5 crayons, 2 cars or 3 dolls, you need to see if they can draw a given number of objects.

For example, ask if your child can draw 2 cats. Can they draw 4 dogs? Can they draw 2 elephants?

Figure 8.4: 4 dogs ?

Once they can draw small numbers of objects you can ask them to draw a larger number of objects.

For example, ask your child to draw 8 dogs, 9 donuts, or 10 mushrooms.

Your child can use crayons, pencils, or bright felt tip markers on colored paper or card. They can use chalk on a chalkboard, or markers on a dry erase board, or a pen on a magnetic drawing screen. These are all readily available at supermarkets or office supply stores.

Watch to see how your child keeps track of how many things they have drawn and how they know when they have drawn enough. Your child will be very happy if you paste the drawing in their diary, or take a photo of their drawing and paste that in the diary.

8.4 Recognition of numbers

Your child also needs to be able to recognize the numbers in their written form: 1, 2, 3, 4, ... A good place to start is at the beginning with the numbers from 1 through 5.

Angela found a packet of brightly-colored magnetic numbers at a supermarket. She tested them and found she could stick them to the refrigerator.

Figure 8.5: Fridge magnet numbers, arranged higgledy piggledy

Angela took the magnetic numbers 1 through 5 out of the packet and placed them in a scrambled order on the refrigerator door, where her daughter Lily could see them.

Angela then asked Lily if she could put the numbers in order, starting from "1".

8.5 Dominoes

Domino pieces are very useful for helping emergent counters recognize and say numbers. Several simple games can be played using domino pieces.

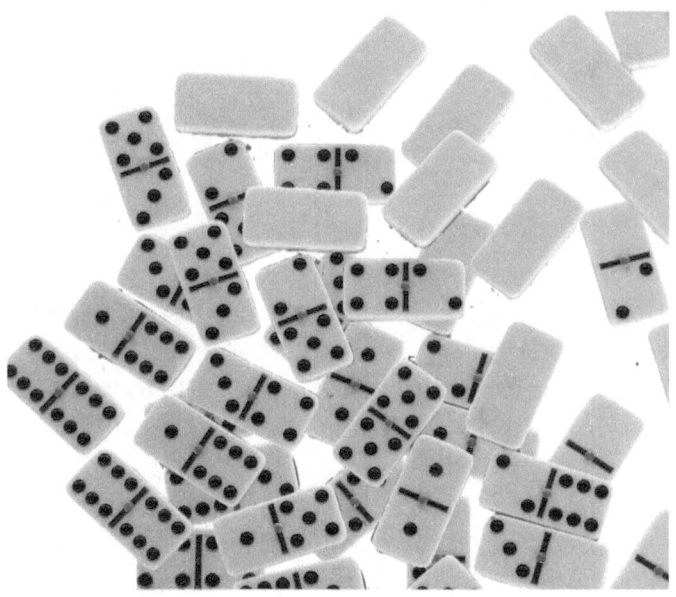

Figure 8.6: Domino pieces

For children who are still learning to recognize numbers and say number names in order, a very simple use of domino pieces is to build domino "trees" in which dominoes are placed one next to another if the number of dots on the abutting sides match:

Figure 8.7: A domino "tree"

This could be a game for one child, where the objective is to build as long or interesting tree as possible. Or it could be a game for two or more children, in which the aim is to be the last player to be able to put down a domino.

These are simple games where physical number recognition, counting to 6, and saying number names, all play a role, helping emergent counters consolidate their number recognition and counting skills.

Chapter 9

Figurative counting

Brian's mother placed some cookies on the table and asked Brian how many there were.

Figure 9.1: Cookies for Brian to count

Brian counted, as he tapped each cookie in turn, and recited:
   "One, two, three, four, five, six."
   "Six," he answered his mom.
   "Six," his mom repeated. "Are you sure?"
   Brian counted again.
   "Yes. Six" he said.

His mother then placed the cookies in a jar. The jar was a solid color and had a lid so Brian could not see the cookies.

Figure 9.2: Cookies in the jar - Brian cannot see them

"See, I've covered them all up," she said. "You didn't eat any while I wasn't looking?" she asked.

Brian laughed. "No!" he said.

Brian's mother picked up one more cookie and placed it in the jar.

"OK. Now tell me how many cookies are in here," she asked, pointing at the cookie jar.

Brian looked at the jar and went to lift the lid.

"No, I don't want you to peek. I want you to tell me how many without looking," she said.

Brian looked at the jar and began moving his hand over the lid, appearing to tap imaginary cookies, while saying "one, two, three, ..."

He stopped moving his hand as he said "six, seven" and then, looking at his mother, said "Seven," with a big smile on his face.

"Seven," she repeated. "But you can't see them. How do you know?"

"Because I counted," said Brian.

But what did Brian count? He couldn't see the cookies. So he could not coordinate saying the number words with touching or pointing at physical cookies.

Brian had previously just counted the cookies for his mom, so he could remember doing that. As he moved his hand over the jar he was mimicking what he would do if he could see the cookies. It seemed liked Brian was seeing images of cookies in his head, though we can't be sure of that.

Brian is a *figurative counter* and he uses his hand movements to stand in for the missing cookies. Brian counts the times he moves his hands over the imaginary cookies.

Brian has internalized his counting. He does not need to see or touch the objects to be counted: he can imagine them, and he can imagine touching them as he says the number names in order.

9.1 What is Figurative Counting?

Figurative counters, like Brian, can successfully answer "How many?" for the tasks of chapter 5. And they answer without guessing. Figurative counters typically do one or more of the following as they count:

- Visualize figures.

- Use body movements such as waving their hands, nodding their heads, tapping their legs softly as they count, biting their lips gently and rhythmically, raising their eyebrows in time with their counting, or rolling their eyes.

- Vocalize counting, as in saying: "seven is 1, eight is 2, nine is 3 ..."

Figurative counters have developed an ability to count hidden objects. However, figurative counters still act as though they were counting physical objects.

A child who is only beginning to be able to count hidden objects will mimic the actions of pointing at, or touching, physical objects as they recite the number names in sequence, just as Brian did.

In other words, beginning figurative counters can count hidden objects but act as if they were seeing those hidden objects in their imagination. Their counting actions look very similar to those of physical counters.

Figurative counting is an ability to count things that a child cannot see, touch or hear. But it is a limited ability. In figurative counting children often use body movements, with their whole body or their hands or eyes, to keep track of counting. As they count, their focus of attention shifts from those unseen objects to the movements of their body.

In the early stages of becoming a figurative counter, a child will usually tap their fingers over hidden objects, as if visualizing them. Later, as they become more practiced, their movements can become very fine and harder to detect. We have seen children tapping their legs softly as they count, biting their lips gently and rhythmically, and raising their eyebrows in time with their counting.

Because figurative counters do not know how many missing objects there are - they cannot see them - they must concentrate hard to keep the rhythmic counting going until they reach the number they know they should.

The name "figurative counting" suggests the children see figures in their heads, but this is not something we can really say for sure.

Children may or may not see images of figures, and they may or may not tell us that they do.

What is essential in this type of counting is that children appear to see images of figures and seem to be rhythmically counting those images as they move their body.

9.2 Counting from one

A very important feature of figurative counting is that figurative counters always start counting from "one" every time they count. Figurative counters cannot start part way through a number sequence.

Alex's mother gave him some crackers and asked: "Alex, how many crackers are there?"

Alex counted: "one, two, three, four, five, six", for the 6 crackers.

Alex's mother gave him 3 more crackers and asked him how many more there were.

He counted; "one, two, three."

Then his mother asked Alex: "How many crackers are there altogether?"

Alex started again from one: "one, two, three, four, five, six, seven, eight, nine."

He could not, yet, begin "seven", "eight", "nine", starting from "six".

Just like Alex, figurative counters must begin again at "one" when they try to count all the crackers.

This appears very similar to children who cannot begin reciting the alphabet from a given letter - say "H". When children first learn to recite the alphabet, saying the letter names in order, they cannot begin the recitation from anything but the very beginning - from the letter "A". As they become more familiar with the alphabet, and more practiced at reciting it, they reach a point at which they can begin reciting from letters other than "A". Figurative counters need to begin counting again from "one", much as children who need to recite the alphabet starting from "A".

9.3 Assessment

Children who are figurative counters can consistently count hidden collections of things. So the tests for physical counters in Chapter 5 also distinguish children who have moved beyond physical counting and are now exhibiting a new form of counting.

Figurative counters can, and do, count collections of things they cannot see, hear or touch.

Children do not develop this ability all at once. There is a transition period where they are unsure of themselves and their counting of hidden collections. They need to check to be sure. But gradually they become more comfortable with counting figuratively and more confident of their answers.

Figurative counters have moved only one step away from physical counting. They no longer need to see or touch items to count them, but they do need to use body movements or mental images to count hidden collections. Children who are figurative counters will use observable body movements in their counting or will appear to be visualizing images in their heads.

Here is a good test to help you decide if a child is a figurative counter. You will need:

- 5 crackers.

- 2 sheets of paper or card, or a cardboard box, or a non-see-through container.

Proceed as follows:

> Place 3 crackers on the table and ask: "How many crackers are there?"
>
> When your child has counted the crackers, place another 2 crackers on the table, away from the first. Point to the 2 crackers and ask: "How many crackers are here?"
>
> When your child has counted the 2 crackers, cover them so they cannot be seen.
>
> Point to the 2 crackers that are covered, and ask: "How many crackers are under here?"
>
> If your child cannot remember, or does not get the number correct, lift the cover and ask them to count again. Then cover the crackers and again ask: "How many crackers are here?"
>
> Now ask your child: "How many crackers are there in all?" Do not let your child look under the cover.

A child who is a figurative counter will be able to correctly find the total number of crackers. However, they will always begin the counting again from "one".

Typically, figurative counters wave their hands over the covers as they count, making rhythmic body movements, and appear to visualize hidden objects.

A variant of this test is to cover both the collection of 3 crackers and the collection of 2 crackers, after your child has counted them, and then ask: "How many crackers are there in all?"

You can do these tests with other small numbers of crackers. But make sure to keep the numbers fairly small, for example 5 or less. The reason is that larger numbers of crackers will stress a child's developing ability to track where they are as they count. Think of this as a sort of mental stress test.

If your child is only a beginning figurative counter you want to start them off slowly where they are more likely to be successful. Then you can gradually increase the size of the numbers so that the task becomes harder for them.

Remember to record what they can do in their counting diary.

You want to give a developing figurative counter the best opportunity possible to show you they can count hidden collections. At this point you are looking for their successes, not their failures, and you want them to show how successful they can be. Also, these small successes encourage your child to continue to show you what they can do.

There are several possible answers your child might give to the hidden crackers task, above:

- They could say they don't know. If this is what happens you need to do more of the assessments for physical counting and help them build mastery of physical counting.

- They might have appeared to have simply guessed. You can check if they guessed by first observing carefully for absence of movements related to counting, and simply asking them: "How did you do that?" A child who guessed will often answer: "I just know!"

- They might make pointing movements over the hidden counters, but get the number wrong. This can happen if their coordination skills - tracking and saying numbers in order - are a little unsure. In this case, get them to do similar examples again with different numbers of counters.

- They might make pointing movements over the hidden counters, counting from "one" through "five" and, looking at the last two fingers used to count, get the number right. In this case they are most likely figurative counters with a little more experience.

- They might count-on from 3: "three, four five". In this case the child is not a figurative counter and has progressed to being able to count-on.

As children develop their counting skills they will be able to answer the question: "How many" for hidden collections like this, but figurative counters will not count-on from a given number bigger than 1.

For example, Michelina, who is five years old, is trying to figure out how many jelly babies are hidden in a jar. She knows there were 6 on the table when she first counted, but now her mom has put some in the jar where they can't be seen.

Figure 9.3: How many jelly babies in the jar?

Michelina is trying to figure out how many. It's hard because she can't see them. She's counted 4 still on the table and as she reaches 4 she says "five, six", nodding her head and tapping her hand each time she speaks. She appears to be visualizing something in her head.
Michelina looks at her mom.

"Two?" she asks.

"You think there's two in the jar?" asks her mom.

Michelina nods her head.

"How did you figure that?" asks her mom.

Michelina is too impatient to answer: she tips the jelly babies out of the jar and counts.

"I was right! Two!" She grins at her mom.

"How did you know?" asks her mom.

"I just counted!" says Michelina.

But what did Michelina count? She couldn't see the jelly babies in the jar, so she couldn't match her counting movements with visual cues from the jelly babies. What was the focus of her attention?

A key is Michelina's hand, head and eye movements as she counts "five, six". She seems to rhythmically move her body as she speaks the counting words, and seems to be looking at a visual image.

Michelina seems not to be focusing on the jelly babies in the jar, which she cannot see anyway, but on how many times she is counting. She seems to know she has to count to 6 because there were 6 jelly babies to start, and none has disappeared or been eaten. She counted to 4 as she matched jelly babies she could see on the table. Now she keeps on counting to 6. Her concentration is considerable and she seems to be looking at something in her mind.

When her mother asks her, Michelina says she was seeing jelly babies in her mind as she counted. Each time she counted - "five", "six" - a new jelly baby would appear as an image. She could see there were 2 of them, and she could feel from her body, head, and hand movements that she had counted 2 times. But she couldn't be sure yet, because she couldn't see the jelly babies in the jar, and she was just learning to count things she couldn't see.

Michelina could not see the hidden jelly-babies but she needed something to count, so she used her hand movements as a stand-in for the jelly-babies she could not see. Even so, she was still not completely sure she was correct.

Michelina has gone beyond the constraints of a physical counter. She has gained the power to count things she cannot. Power indeed! She has moved to figurative counting.

Importantly, when counting to find the total of two collections - say 6 crackers and 3 crackers - figurative counters usually count both collections, always starting from "one" for each separate collection, and then always counting the entire collection, again starting from "one".

Figurative counters are not yet able to count-on from the result of a previous count. A figurative counter cannot begin with "six" and say "seven, eight, nine" in order to find the total of 6 and 3 crackers.

This may seem odd to you as an adult - isn't it obvious to a child that "seven" comes after "six"? Yes it is, but understanding that "seven" comes after "six" is not what prevents a child from counting-on. The heart of the matter is what "seven" - or any other number - stands for in a child's mind and experience. Does it stand for the number after "six" or does it stand for a memory of having counted "one, two, three, four, five, six"?

We will pick up this point again in chapter 10 when we look more closely at children who can count-on. For now you should bear in mind that if your child cannot count-on there is a very strong likelihood that, at this time, they are a figurative or a physical counter. The various tests we have described in this chapter will help you decide if that it is the case.

Tests for physical counters are good tests for figurative counters. The big difference is that figurative counters can count hidden collections. But being able to count hidden collections does not in itself tell us that a child is still a figurative counter: they may, in fact, have progressed to another more efficient type of counting.

Initially, when children combine two separate collections, such as one with 6 counters, and another with 3 counters, the child does a triple count. For example, to add 6 counters and 3 more counters that have been hidden the child counts the 6 counters: 1, 2, 3, 4, 5, 6; then the 3 counters: 1, 2, 3 and then finally the whole collection of 6 and 3: 1, 2, 3, 4, 5, 6, 7, 8, 9.

As they become more efficient, the children still start from 1 and still make some sort of movements with head, hands or eyes to track the counters they can see - in this case the six counters - and the ones that are covered: the three counters under the cover. Counting by ones starting at one is often called "counting all". The children are still not counting on.

One task we have used with young children highlights the different strategies used by figurative counters. You will need:

- 9 counters
- 1 piece of card.

Proceed as follows:

> Place 6 counters on the table in front of the child. Point to the counters and say: "Can you count the counters for me?"
>
> After your child has counted the 6 counters place 3 more counters under a piece of card and say: "Under this card I have 3 more counters. How many counters are there in all?"

There are two different types of successful responses to this task. Some children count the 6 counters again: 1, 2, 3, 4, 5, 6. They then tap over the card three times: 1, 2, 3. Then they re-count the six counters and the three taps: 1, 2, 3, 4, 5, 6, 7, 8, 9.

Other children recount the six counters but keep the count going as they count the three taps or nods over the card: 1, 2, 3, 4, 5, 6, 7, 8, 9.

In his original research studies, Leslie Steffe used a variant of these tests. For this task you will need about 10 counters or objects. You proceed as follows:

> Place the counters (or objects) on the table.
>
> Ask the child to count how many counters there are.
>
> After the child has counted, asked them to look away while you remove some of the counters from the table - say 4 - and place them under a card or cloth.
>
> Then say: "How many counters can you see?"
>
> After the child counts the counters ask: "How many counters did we start with?"
>
> Then ask: "How many counters are under the card (or cloth)?"

Again, physical counters cannot answer this question. Some beginning figurative counters will not be able to answer this question either. If your child can answer this question correctly, it is a very strong indication that they have moved beyond physical counting to figurative counting, and perhaps even beyond that. If they can answer this question but not count-on - which we discuss in the next chapter - then they are definitely a figurative counter.

As children become adept at figurative counting they are able to answer more difficult questions that a physical counter could not answer.

Bobby is an 8 year old figurative counter trying to figure out 7 + 9. First Bobby says he sees 7 blue balls, numbered 1 through 7, in the left side of his mind.

Then he sees 9 blue balls numbered 1 through 9 to the right. He imagines the ball numbered to the right numbered "9" moving to the left, where it gets re-numbered "8".

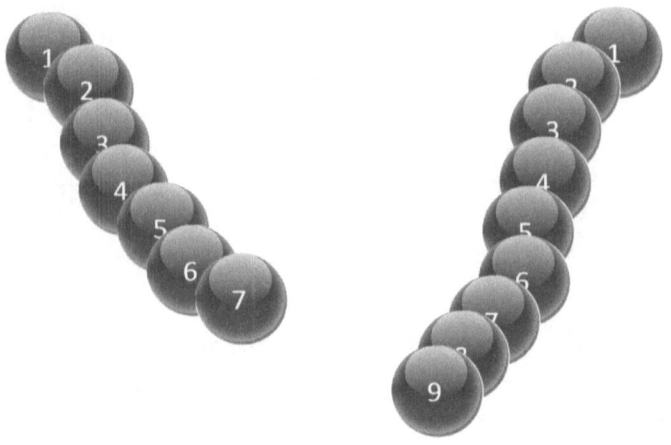

Figure 9.4: Bobby begins to visualize 7+9

Now the ball to the right numbered "8" moves to the left where it gets re-numbered "9". Then the ball to the right numbered "7" moves to the left and becomes "10".

With deep concentration, his eyes moving as he does it, Bobby gets an answer of 16. He looks at his dad, waiting for a sign that he's correct.

Bobby is pleased when his dad agrees that 7 + 9 is 16, but Bobby sighs as he wonders if he'll ever be able to do sums like this as quick as some of the other kids. He wonders what they see when they do addition sums and how they can count so quickly.

A child who is a figurative counter has made a big step beyond being a physical counter. They now have an amazing power that a physical counter cannot even imagine: they can count things they cannot see, hear, or touch. This is a considerable intellectual achievement.

Yet, as Bobby's working out of 7 + 9 shows, figurative counting can sometimes be hard work. It can put a big strain on working memory, the aspect of our brains that deals with holding and using information in real time.

A figurative counter's activities have become internalized as they focus on their body movements, and keep in mind their counting acts. With practice, figurative counters can get good at internal counting. So good, in fact, that they do not want, or do not see a need, to give it up. Why should they? Figurative counting is successful for them and allows them to do things that, as physical counters, they could not do.

The problem is their working memory: it's capacity is limited and it seems that no amount of practice can significantly alter how much working memory a person has available. So the only way to overcome the problems Bobby had with those hard addition sums is to adopt a different approach, a different way of thinking about numbers and counting.

9.4 Meaning of numbers for figurative counters

A figurative counter has different memories of counting than does a physical counter.

A physical counter remembers acts of counting, of establishing one-to-one correspondence by saying number names, in correct order, as they focus on physical objects that they can see, feel, or hear.

A physical counter's focus of attention is on saying the number names and matching them one-to-one with physical objects and then taking the last number they say as the answer to the question: "How many?"

In contrast, a figurative counter has a different focus of attention. Their attention is focused on trying to imagine hidden objects that they cannot see, and then counting those hidden objects.

Because this is a hard task for them, they will most likely use a physical stand-in for the hidden objects. Commonly these physical stand-ins are fingers. As they become more efficient, physical counters will often use body movements such as tapping, jaw movements, or raising eyebrows, as stand-ins for imagined objects. A figurative counter then physically counts these stand-ins.

Because figurative counters work at establishing physical stand-ins for hidden objects, their memories of counting are different to those of a physical counter. As they count more and more, using figurative counting, they build up skills of establishing very fine physical stand-ins, such as very, very light tapping, as they count imagined objects, and their memory gradually shifts away from the physical stand-ins toward the imagined objects.

Gradually, therefore, figurative counters begin to imagine a certain number of hidden objects, without yet knowing what that number is, but knowing they can work it out using physical stand-ins. Their memories become more and more those of internal imagined processes. The process of counting is becoming more internalized as a thought process.

Chapter 10

Counting-on

10.1 Counting-on versus counting-all

Five year old Tim does not always need to count collections of things by starting from "one".

He is able to solve the following problem that his dad gave him: "Tim, can you see 8 crackers on the table? How many more would you need to have 13 crackers?"

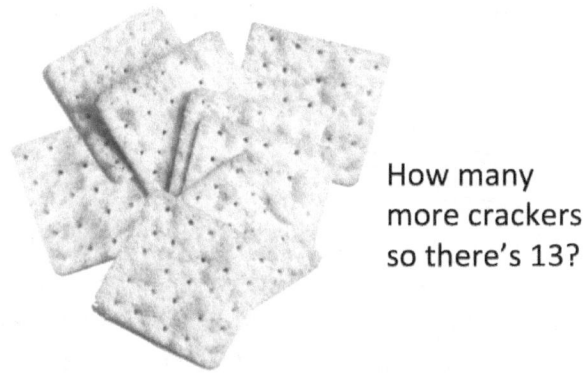

How many more crackers so there's 13?

Figure 10.1: 8 crackers - how many more for 13?

Tim solves this problem by counting on from 8: "nine, ten, eleven, twelve, thirteen". As he counts, to solve this problem Tim moves his jaw from side to side.

Tim's dad thinks Tim counted how many times he moved his jaw as he said "nine, ten, eleven, twelve, thirteen", and so concluded that 13 is 5 more than 8.

Tim is at the very beginning stages of learning how to count-on. He uses body movements, coordinated with saying a number that does not begin at "one". He counts his body movements as he does this. Tim has to focus his attention very carefully and coordinate two sets of movements: his speech and his jaw movements.

In one important sense Tim seems like a physical counter: he counts the movements of his jaw. But Tim is doing much more than that. He begins saying part of a number sequence, beginning at "nine" - the point he would have been at if he had counted from "one" through "eight". This indicates to us that he can extract a number like "eight" from the sequence of counting numbers and begin his counting from there.

Tim can begin saying counting numbers from points other than "one". Tim is at the beginning of being able to count-on.

Children who can count-on - that is, begin counting from a given number - will usually use this ability to figure out a problem such as "8 + ? = 13", much as five year old Tim did.

They will also use counting-on to solve addition problems. For example, a child who can count-on will typically answer a question such as "What is 7 and 4 more?" by counting-on from "seven": "eight", "nine", "ten", "eleven".

At first, children count-on to solve addition problems by starting from the first number presented to them: for 3 + 7 they will typically begin by counting-on from 3. Only later, as they have mastered this aspect of counting-on, will they begin by counting-on from the larger number (7, in this example.)

It sometimes seems that children who are only just beginning to count-on are still focusing on images of objects to be counted, such as marbles, counters, or jelly babies.

However we see with the example of Tim, and other children who tap their fingers when they count-on, that sometimes children who count-on are focusing mainly on body actions. That is, the children make body actions such as tapping, or biting their lips as they count-on.

It may be that they do not have visual or other concrete images in their minds. Rather, they may be focusing on keeping in working memory the number of taps or bites they are making. Whichever is the case - focusing on mental images or on real-time body movements - the thing these children are not doing is taking the number names "one", "two", "three" ... as things to count.

Children who count-on have made a huge advance beyond figurative counting, They are flexible with regard to where their counting begins.

10.2 Counting-on & meaning of numbers

When a child can count-on, a number such as "eight" has two meanings. First, it is the number in sequence immediately following "seven". Second, it is now a trigger for a memory of having counted "one, two, three, four, five, six, seven, eight".

For a child who can count-on, a number name such as "eight" brings to mind these two sorts of memories. The first is a memory of "eight" as following "seven". The second is a memory of "eight" as being the last name spoken in a sequence beginning with "one".

The long-term memory of previous counting and its use in counting-on is a sign of mental flexibility in a child's development.

However, the sorts of things such children can count are still what figurative counters count: objects, images, bodily movements, or memories of these things.

Children who can count-on have a more advanced reference for number names such as "seven" and "eleven".

When a child is an emergent counter they are just learning the number names, learning to sequence them in the order they hear from their parents or care-givers, and learning to match up actions with number names.

Children who can count-on have come a long way from being an emerging counter. They have progressed though physical counting, and being able to count small hidden collections of objects. Numbers are now beginning to take on a more symbolic meaning for them.

10.3 Parting of the ways

A child's ability to count-on - or not - is so important for their mathematical growth that Eddie Gray, then at the University of Warwick in the UK, identified this point in a child's potential counting development as "the parting of the ways."

Figure 10.2: Eddie Gray

What he meant by this is that the use of counting-on by a child shows a marked degree of flexible thinking about numbers, and that flexibility of thought allows further rapid development in children's understanding of, and proficiency with, number.

The question for us, and for you as parent or care-giver, is what we can do to help stimulate and develop the flexibility of thought involved in counting-on.

Ultimately it is a question of memory and focus of attention, which is where you, as parent or care-giver, can really help.

Here is a simple exercise to see if your child is able to recall and focus in a way that allows them to begin to count-on:

> Ask your child to count to eight, by ones, starting from 1.
>
> When they've done that say: "Good! Now can you tell me what is the number just before eight?"

Usually one of three things will happen:

1. Your child will say "seven".

2. Your child will say a number other than 7.

3. Your child will not know and cannot say.

In case 1 your child is showing an awareness of the number just before another number. You can help them build on this memory and focus of attention by playing a game: what comes just before and what comes just after. To play this game you and your child take turns in quizzing each other as to what number comes just before, or what number comes just after, different numbers, for example, you may ask:

"What number comes just before 12?"

"What number comes just after 14?"

Alternate with your child asking *you* similar questions.

Watch carefully to see if your child seems to be counting up from 1: for example if you asked what comes just before 12, does your child answer "11" without apparent counting, or do they seem to be counting up from 1?

The aim of this game is to focus attention on numbers just before and just after a given number and to form lasting memories of those facts.

It is just this focus of attention and development of memory for facts that will allow your child to begin to count-on, and to develop a greater flexibility in using numbers.

The same game can usefully be played with children who answered as in cases 2 and 3, however you will want to take it slowly and give the child lots of chances for success by starting with smaller numbers well within their capabilities. For example:

"What number comes just before 4?"

"What number comes just after 5?"

Again, watch carefully to see if your child seems to be counting up from 1. If they do seem to be consistently counting up from 1, encourage them to guess rapidly:

"Can you guess, as quickly as you can, what number comes just before 7?"

If they answer anything other than "6" get them to count from 1 and focus attention on 6 being just before 7.

Playing the game *Chutes and Ladders* (also known as *Snakes and Ladders*) will highlight if your child has to count by ones (that is, are figurative counters) or can count on.

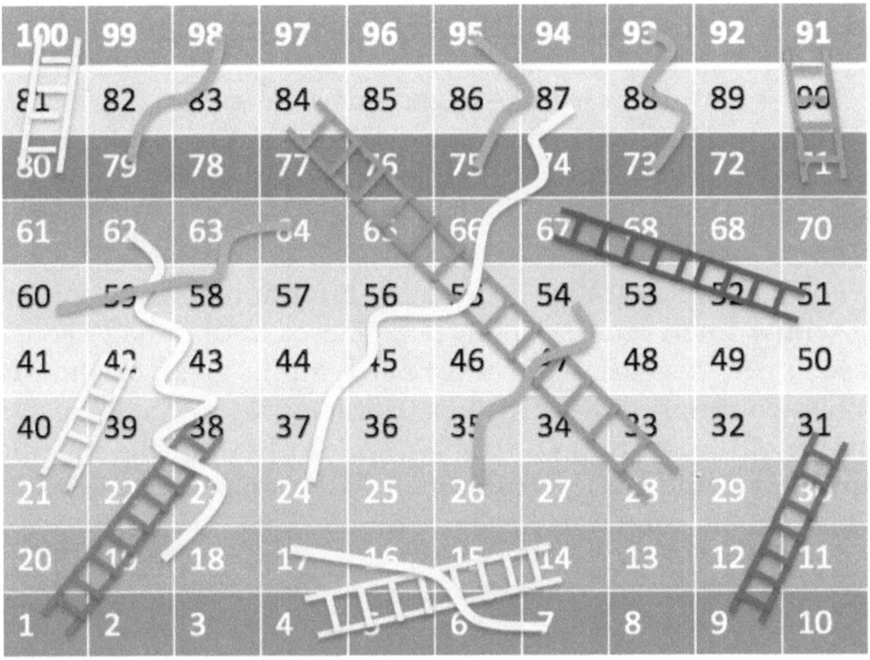

Figure 10.3: Chutes and Ladders

Whether you use a spinner or a die to determine how many moves each player should move forward, you can check if your child, or children, tap each square on the board once or simply move forward the appropriate number of steps. For example, if a child's token is on square 23 and they spin, or throw, a 4, does the child tap the squares beyond where they are: 24, 25, 26, 27, or nod their head rhythmically as they look at each square in turn, or do they simply move forward 4 spaces without apparently counting by 1's? If it's the latter then this may be evidence that they are counting-on from 23.

Remember, the ability to count-on relies on a simple yet powerful form of mental flexibility: the ability to see a number such as "eight" *both* as the number in sequence immediately following "seven", *and* as a trigger for a memory of having counted "one, two, three, four, five, six, seven, eight".

10.4 You cannot "teach" counting-on

This is a health warning - a cautionary note.

Many teachers and adult helpers believe that all children in the early grades of school should be able to count-on, and for those who cannot, the teacher should show them how it is done, by direct instruction.

For example, a grade one teaching assistant, Marie, was working with a small group of children. These children had been assessed by the teacher as the most needy in the class, on the basis of their counting skills. The task for them was to choose a small number of blocks from a container, and say how much more than 4 the total would be. The children had all been given a work-sheet with numbers and totals to be filled in:

$$4 + \square =$$
$$4 + \square =$$
$$4 + \square =$$

Figure 10.4: $4 + ... = ?$

Edward, a 5 year old, reached into the container and picked out 5 blocks.

Marie asked him to write 5 in the first box.

Marie already knew Edward would not be able to count the 5 blocks and 4 more without seeing the other 4, so she asked Edward to take out 4 more blocks. She then asked him how many there were in total.

He bunched all blocks together and counted: "one, two, three, four, five, six, seven, eight, nine" He then wrote 9 after the equality sign.

Edward repeated the exercise, this time choosing 3 blocks from the container. Again he bunched all the blocks together and counted from 1 through 7, before writing his answer.

At that point Marie said to him:

"Edward, you are always counting 4 more than what you choose, so just start from 4."

"OK", said Edward.

He reached in the container, and took out 4 blocks. He then took out 4 more, bunched them together, and counted from 1 through 8.

Marie stopped him.

"Edward, did you listen to what I said? You can begin counting from 4. Like this: four, five six, seven, eight."

Edward nodded his head, reached into the container and took out 8 blocks. He took out 4 more, bunched them together, and counted from 1 through 12.

Marie shook her head in frustration.

Of course Marie was frustrated: to her, as an adult, it just made sense to count from 4 when you always had 4 to add to the new number of blocks from the container. All she was doing was showing Edward how to do that.

Sadly, for both Edward and Marie, Edward was not yet able to count-on. Edward needed work on counting physical objects, and then help progressing to being able to count hidden objects he could not see, such as crackers in a tin:

Figure 10.5: Counting crackers

Edward first needs to learn to figure problems like: there are 2 crackers on the table and 4 more in the tin. How many are there in all?

No amount of telling, in that class period, was going to help him count-on.

Careful assessment of a child is of the utmost importance if an adult is to help that child progress to more efficient methods of counting.

A child's self-esteem rises when an adult takes interest in what they can do, listens carefully, assesses thoughtfully, and then suggests new things for the child to do, at which they can be successful.

Edward was in a situation where Marie, meaning well, was rapidly lowering Edward's self-confidence with counting, because he could not yet do what Marie seemed to say he should be able to do.

Assessment, dear parents, is critical to helping your children.

Once we assess Edward as a physical counter we need to enhance his physical counting skills to the point where he is very fluent with those skills. Then, and only then, do we begin figurative counting tasks with him. Edward has some way to go before he can fluently count-on, but with careful assessment, and focus on mastery of his existing counting skills, he will get there - and probably sooner than we imagine.

Chapter 11

Helping physical counters

11.1 Transition to figurative counting

If you have assessed your child as a physical counter the next step is to help them become a figurative counter - a child who can count hidden objects.

Alicia has assessed her son Jon and found that he has difficulty counting hidden objects. She feels reasonably sure that Jon is still a physical counter.

What can Alicia do to help Jon progress to moving beyond physical counting to become a figurative counter, to develop an ability to count hidden objects?

The idea is to start with activities that a figurative counter can do that a physical counter cannot: counting hidden objects.

"Wait!" you hear Alicia say. "Jon is a physical counter! What good is it going to do to give him hidden objects to count? I already know he cannot do that. Aren't I just setting him up to fail if I ask him to try?"

Yes, Alicia would be if she made the task too hard, which we know it most often is for physical counters.

The process that we have found works is to make the hidden

counting task as simple as possible. This is because we do not want Alicia to over-burden Jon's working memory.

Alicia knows Jon is a good physical counter. Her problem now is the following:

How to think of something that will stimulate Jon to think about the *results* of his physical counting, rather than just have him focus on the *process* of physical counting?

Alicia understands the idea of learning through success. Jon needs to reflect on his actions of physical counting, to be less focused on carrying out physical counting - which he does very well - and more focused on the outcome of that counting.

To begin, we will ask Alicia to use a very simple starting place, where she can help Jon have success at counting things he cannot see or touch.

Alicia gave Jon 2 chips and asked him to count them. This is a very small number, completely within the counting range of a physical counter.

Now she picked up pick up one more chip: just one, no more.

Alicia showed this extra chip to Jon, and said: "Here's one more chip. I'm going to place it under this mug."

Then she placed the single chip in her hand under a non-see-through mug, and asked Jon: "How many chips are there altogether?"

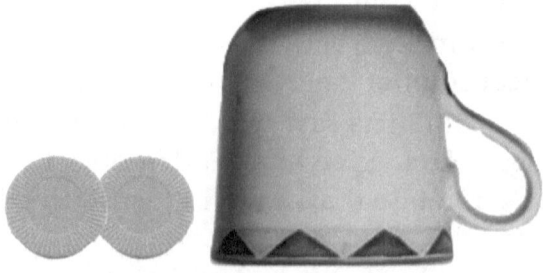

Figure 11.1: 2 chips on the table, 1 under the mug: how many in all?

Either Jon will not know, or will say "three".

In fact, Jon did say "three".

Alicia then asked him: "So you think there's three chips in all?"

Jon nodded affirmatively, and said: "Yes".

Alicia then asked Jon: "Shall we take a look?"

"Yes", he said, so Alicia asked him to lift the mug and count the chips.

When Jon counted "one, two three", Alicia said: "Three! You were right! Wow! You sure are a good counter!"

Figure 11.2: Jon was right!

Alicia has scored a hit! She got Jon to think about his counting, by placing just a single counter under the mug. She only used 3 counters in all, because she knows that physically counting only 3 counters is way easy for Jon.

Alicia will reinforce Jon's first steps at figurative counting with more counting tasks like the one she just gave him.

If you try this and your child does not know there are 3 chips in all, try doing the exercise over, starting with just one chip on the table and one under the mug.

If your child cannot figure out there are 2 chips in all, then make a note in your counting diary, relax, and go back to practicing physical counting.

The time will come, very soon, when they can complete this hidden counting task.

There are two ways to extend this task.

First, you can increase the number of chips you give your child to physically count.

Second, you can increase the number of chips you place under the mug, out of sight and touch.

In our experience it is a good idea to mix these two tasks, randomly increasing the number of chips you give to your child, or place under the mug.

The reason is, simply, that you want your child to become versatile at counting hidden objects.

If you only increase the number of chips you give them, in order, from 2 to 3, then 4, 5, 6 and so on, while always having only 1 chip under the mug, your child might spot a regularity in what you are doing, and so anticipate the next answer, without actually being able to count hidden objects.

Mix it up a little, while still keeping the tasks very simple.

For example, give you child 3 chips and place 1 chip under the mug.

Then, when they can count how many there are, give them 2 chips and place 2 under the mug.

Alternate increasing the number you give them, and the number you hide under the mug.

When they are showing a lot of success, try increasing the number you give them, or hide under the mug, by 2 or by 3.

Always emphasize success, and make a diary note of what they can do.

11.2 What to look for

A key feature to look for as children make a transition from physical counting to figurative counting is that they begin to be able to count some objects they cannot see, touch, or hear. They will begin to do this from time to time, but not yet regularly. These occasions are critical, and you should note them in your child's counting diary.

Typically, as children begin to be able to occasionally count hidden items, they will make physical substitutes for the imagined objects they are trying to count.

A common thing children do at this transition is to use finger patterns to stand-in for the hidden objects they are trying to imagine.

Children then do a physical count of the fingers they see. You can see clearly that a child at this point of moving from a physical to figurative counter is using (a) what they know to do - physical counting - and (b) what they have - their fingers - to solve the problem of counting hidden objects.

Something to watch for during this transition point is that a child in transition from physical to figurative counting will sometimes lose track of their counting. They might not make a correct one-to-one correspondence.

This is because they are carrying out a task that is difficult for them: using fingers to stand in for imagined objects and, at the same time, trying to count those fingers as they go.

This is like a mental stress test for young children, so sometimes, when they wobble a little under stress, it becomes clear to you as an observer.

Make a careful note of any of these counting slips in your child's counting diary: they are a very good indicator that a stable physical counter is in transition to figurative counting.

11.3 Further activities

Another activity that helps many physical counters move to figurative counting is placing chips into a container that your child cannot see into, such as under a mug, or in a piggy bank.

Figure 11.3: Counters in a piggy bank are out of sight

Ask your child to count out loud as the chips go into the container.

Then show them one more chip and ask how many there are altogether.

This is a hidden counting task that is sort of the reverse of placing few extra chips in a mug. In this task your child counts chips they can see and touch, and the chips are then placed, one by one, out of sight. They then see the extra chips you give them, and have to figure out how many there are in all. What's hidden in this task is what they have already counted.

Again, keep this simple starting with just 2 chips to count and place out of sight, under a mug, or in a piggy bank.

Then give them just 1 more chip and ask how many there are in all.

The same principle of gradually increasing the number to count and hide, and the number you give them, applies as before.

Joshua was helping his daughter Rachel to count hidden objects.

Joshua said: "Rachel, pretend there are 3 counters in the tin. Here are 2 more counters," and he showed Rachel the 2 extra counters.

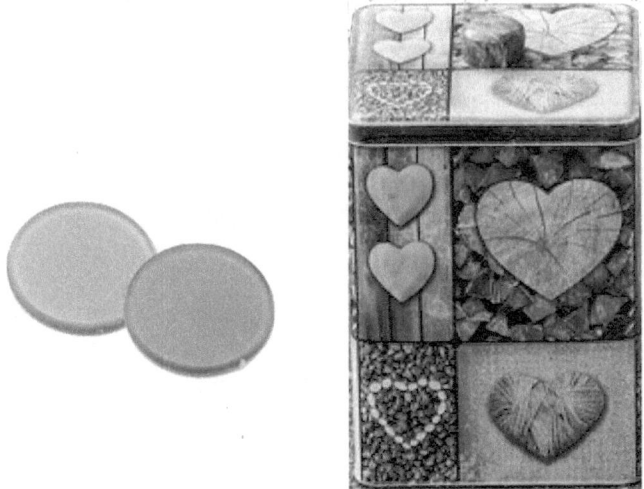

Figure 11.4: 3 counters in the tin - how many in all?

He then asked Rachel: "How many counters do I have in all?"

Rachel could answer 5 right away, so Joshua said to her: "Pretend I have 10 counters in the tin and pretend I have 4 in my hand. How many do I have now?"

Rachel thought a lot harder about this question and asked her dad to say again how many were in the tin and how many in his hand, before answering "14".

Chapter 12

Helping figurative counters

12.1 The burden of counting-all

Lara is a figurative counter.

Lara's mother showed her two dice. One had 6 dots showing face up, and the other 3 dots showing face up.

Figure 12.1: Lara's mother picks up the dice and asks Lara the total

Lara's mother asked her how many dots there were on each.

Lara looked at the dots and said "six" and then said "three."

So she recognized and named the pattern of dots.

When her mother picked up the dice, wrapped her hand around them so they weren't visible, and asked Lara to count how many dots there were altogether on the two dice, Lara tapped on her face with her fingers as she seemed to track the dots she had seen on the dice.

For each dot she counted Lara used a different finger moving from thumb to small finger on one hand for every five she counted. She tapped from thumb to small finger then thumb again for a count of six. To count the number of dots on the two dice she first counted the six dots (tap, 1,..., 6), then the three dots (tap, 1, 2, 3) and then tried to replicate the taps (1 - 9) moving in order from thumb to small finger as many times as she needed. Lara had trouble with the last count because it was difficult for her to keep track of the number of things she was counting.

Because she tapped once for each of the 6 dots, once for each of the 3 dots, and then once again for each of them combined, Lara had to make 18 taps to represent the total of 9 dots. This meant it was easy for her to lose place.

Sometimes when she does counting like this with her mother, Lara does not tap her face with her fingers, but nods her head, blinks or moves her eyes, taps her leg with her hand, or taps under a table.

12.2 Progressing in figurative counting

Children, like Lara above, often go through a progression within the figurative stage of counting.

The first aspect of this progression is when, for example, children are asked to add 6 and 3 they will often, like Lara, count to 6 starting at 1 then count to 3 starting at 1 then attempt to count both collections - the combined 6 and 3 - as one group, starting at 1.

Often they have limited success in doing this, because, like Lara, they can easily lose track of where they are. They count each collection separately, often without success, because their co-ordination seems to get less sure when they try to count the combined collection. Success, of course, depends on being able to make one-to-one co-ordination, and a firm knowledge of the verbal number sequence.

The second step of the progression through figurative counting occurs when children can successfully count the three collections separately because they know the verbal sequence and have one-to-one correspondence firmly in place.

An efficient step of the progression through figurative counting occurs when a child knows that the two collections need to be seen as a single entity and can count this collection, starting at 1, in one count.

Our experience is that unless a child can successfully count the two collections as one whole collection then there is very little chance that they can even begin to count-on.

At what part of the figurative stage is a figurative counter? Here are some important questions for you to answer:

- Do they count each collection separately but incorrectly?

- Do they have one-to-one correspondence? Evidence of one-to-one-correspondence?

- Do they have the verbal sequence in place?

- Do they know that the last number name spoken for the count is the number in the set? Count each collection separately and correctly?

If a child is a figurative counter and you are wondering if they are ready to move to counting-on you need to check if they can succeed easily at the following typical tasks:

- Give your child 5 counters and ask the child to count them. Hide one counter under a piece of paper and ask: "How many are there now?" Repeat this scenario numerous times using various starting numbers before then hiding two counters, then three, and so on.

- Place counters into a container that your child cannot see into, a piggy bank for example. The child counts out loud as the counters go into the container and then you show them one more counter and ask how many there are altogether.

- Say to your child:

    - "Pretend there are 3 counters in the tin. Here are 2 more (*Show them the counters*). How many do I have altogether?"

    - "Pretend I have 10 counters in the tin and pretend I have 4 in my hand. How many do I have now?"

If you feel your child is at the early stages of figurative counting and not ready yet to move to counting-on, it will help them if you focus on mastery of the basics.

Practice with verbal counting, asking your child questions such as: "What number comes after? Before? Is one more? Is one less?"

Your child might have more difficulty giving you the number before a number than the number after. You can check how they work this out. For example, when asked to give the number before 9, some children need to count forwards from 1 through 9 then try to remember the number they said just before 9 before they can answer.

When children are confident with the verbal sequences starting at 1 choose a different starting and finishing point within the range 1 to 20.

For example:

- Give your child a piece of paper with numbers 1 through 20 written on it.

- Ask your child to put their finger on a given number, say 5, then say all the numbers stopping at 17 (for example).

- It is important that your child initially touch all the numbers as they say them. These numbers can be varied and can include counting backwards. For example, start at 19 and count backwards by ones. Stop at 3.

- Can your child also count forwards and backwards by ones correctly without touching each of the numbers?

To assist figurative counters to move to counting-on we need to ensure that they can order both consecutive and non-consecutive numbers starting at a number other than 1.

For example, given the numbers from 6 to 12 can they place them in order from smallest to largest and largest to smallest?

Given a random collection of numbers, for example: 4, 9, 1, 14, 20, 3 can they place them in order from smallest to largest and largest to smallest?

12.3 Moving to counting-on

When you feel your child has a behavioral mastery of figurative counting, and it's time to help stimulate them to move to counting-on, what sort of activities are productive?

First, we are using a child's mastery of figurative counting to help them learn how to count-on. This means they should be fluent and capable with figurative counting - especially counting of hidden things.

Second, a critical aspect of the ability to count-on is for a child to be able to see a number, such as 9, as two things:

1. A memory of the *process* of counting by 1's: 1, 2, 3, 4, 5, 6, 7, 8, 9.

2. A separate memory of the *result* , namely "9" of counting by 1's from 1 through 9.

As adults we make think these two things are one and the same and are obvious. To a child who cannot yet count-on they are not the same and they are not obvious.

The issue for us as adults is to help our children focus awareness on the *result*, and not only on the *process*, of counting to a particular number.

Remember the story, in Chapter 5.4, of five-year old Annie, a physical counter, who was writing how many people would fit in the seats of a drawing of a cinema. She had reached 27 when an older friend, Jamie, asked her: "Which number are you up to, Annie?"

Annie answered: "Twenty seven."

Jamie then asked: "How many people would fit in those seats you've written on?"

Annie counted, from the beginning: "One, two, three,..., twenty-six, twenty-seven."

"Twenty-seven", she said.

Annie could answer the "How many?" question by counting from one, using one-to-one correspondence, but she was not aware that the number "27", which she herself had just written, was also the answer to the "How many?" question. Annie's meaning for numbers did not yet extend to knowing that a number name is the number of objects there are in a collection.

Annie was a physical counter and not yet a figurative counter - she could not yet count hidden collections of things. But even figurative counters can be stumped, as was Annie, and not be aware that a result they have just obtained through counting is an answer to the question "How many?"

Let's see how this might play out, in an example.

Five year-old Rory was doing some counting exercises with his dad Peter. Rory loved these games and was keen for his dad to give him more counting problems.

Peter showed Rory a picture of some geese outside a shed, and asked Rory: "Rory, how many geese do you see?"

Figure 12.2: Geese outside a shed

Rory looked carefully and said: "Five!"

"OK!" said his dad. "Great! Now there's another three geese in the shed. Did you know that?"

"No", said Rory.

"Well there are - three more geese in the shed. Do you know how many geese that makes altogether?"

"Wait, wait!" said Rory, and he counted on his fingers: "1, 2, 3, 4, 5" and then kept counting, while putting up another 3 fingers: "6, 7, 8. Eight!"

Rory's dad had been watching and he noticed that Rory began counting at 1, and when he reached "5", for the five geese that he could see in the picture, he put up three more fingers in turn and counted: 6, 7, 8 ". At no point, however, did Rory look like he was going to simply count "6, 7, 8", starting from "5".

Rory's actions indicate (but do not prove conclusively) that he was focused on the *process* of counting from 1 through 5 and not on the *result*, 5, of that process.

Peter was aware of the distinction between a counting process and the memory of the result of that process, so he tried a few more hidden counting scenarios with Rory and found that Rory consistently counted from 1.

Peter's conclusion was that Rory was not yet able to count-on.

However, Rory's figurative counting abilities were very strong and consistent as Peter noticed when Rory counted the missing geese: he didn't have to count the five geese, then the 3 missing geese, then the total collection. Rory still counted from 1, but he was able to continue counting by 1 when he raised the extra 3 fingers, standing in for the 3 geese in the shed.

Peter wondered what sort of things he might do to stimulate Rory to begin counting-on.

Because Rory seemed to have behavioral mastery of figurative counting, Peter decided to play a slightly different game, one that might encourage Rory to focus on the number of things present, rather than on the action of counting them by 1s.

Figure 12.3: Beach chairs

Peter said: "Rory, how many beach chairs do you see in this picture?"

"Three", answered Rory.

"Are you sure?" asked Peter.

"Yes! Look: 1, 2, 3. See."

"OK, good" said Peter. "Now some people came along and one sat in each chair. How many people were there in the chairs?"

"Three", said Rory, without hesitation.

"How do you know?" Peter asked.

"Because it's the same as the chairs", said Rory.

"OK, same as the chairs. Well then each person sitting in the chairs reached in their bags and pulled out a bottle of soda. How many bottles of soda were there?"

"Same as the people - three", said Rory.

"Good", said Peter. "Now someone brought along three more chairs. How many chairs are there altogether?"

"Six", replied Rory, "see: 3, then 4, 5, 6."

"Excellent!" exclaimed Peter, and he was indeed very happy because Rory had just taken a first step to counting-on.

By focusing Rory's attention on the *result* of the count to 3, namely the number "3" itself, and away from the *process* of counting from 1, Peter had allowed Rory's brain, already free from much of the strain of counting figuratively, to use that focus to succeed in counting-on to 6.

Peter emphasized in several ways, the number 3, which was, of course, the number of chairs in the picture. Peter's aim was to focus Rory on the number of chairs there and not on the process of counting "1", "2", "3".

Peter deliberately chose a relatively small number - 3 - so as not to put too much stress on Rory's brain as he focused on 3 more chairs coming into the picture.

Makayla had been thinking about how she might help her daughter Alyssa move to counting-on. Makayla had been watching Alyssa's counting development and knew she was a strong figurative counter. It's not like Makalya wanted to say: "Now listen up Alyssa: I'm going to show you how to count on." Makayla knew she had to invent a game for Alyssa that would encourage her to begin to count-on, building on her strong figurative counting skills.

Makalya found a piggy bank and a bunch of counting chips. Alyssa looked interested to see what her mom was going to do.

Makayla asked Alyssa to count out 6 chips:

Figure 12.4: The piggy bank and 6 counting chips

"OK, so how many chips are there?" asked Makayla.
"Six", said Alyssa.
"Six", repeated Makayla. "Not five?"
"No, six!" exclaimed Alyssa.
"Are you sure?" asked Makayla. "I think maybe there's four."
"No, there's six. Look: 1, 2, 3, 4, 5, 6."
"OK. Six chips. Can you put those six chips in the piggy bank for me?"
Alyssa placed the chips in the piggy bank and Makayla asked:" How many chips did you put in the piggy bank?"
"Six!" exclaimed Alyssa.
"Not four?" asked Makayla.
"No!! I put in six!" said Alyssa.
"I guess it was six", said Makayla. "You know what, I'm going to put in three more."
Makayla picked up three chips and said: "1, 2, 3" as she put

them in the piggy bank.

"How many chips are in there now?" Makayla asked Alysaa.

Alyssa looked for a couple of seconds and said "Nine!"

"Nine", echoed Makayla. "How do you figure that?"

"Well, look: it's 6 - 7, 8, 9", said Alyssa as she raised three fingers in turn after saying "6".

"Let's see if you're right", said Makayla, opening up the piggy bank and emptying the chips.

Alyssa counted nine chips and laughed to her mom: "I was right!"

Makayla put a strong emphasis on the number of chips - 6 - that were going to go into the piggy bank. She wanted to focus Alyssa's attention away from an act of counting by 1's to noticing that "6" stood for the number of chips laid out next to the piggy bank. She focused on the number 6 by playing around with numbers 4 and 5 that are close to 6, getting Alyssa to emphasize there were 6 chips on the table that went into the piggy bank. This focus on the number 6, away from the act of counting 1, 2, 3, 4, 5, 6, was enough of a scaffold to allow Alyssa to start from 6 and simply count 7, 8, 9.

Chapter 13

Consolidating counting-on

13.1 Miscounting by 1

When your child starts to use the count on strategy consistently to solve addition tasks their response is sometimes one number less than the correct one. Usually this is because they include the number from which they should be counting-on.

For example, if you ask your child how many cookies there would be if you had 6 cookies on one plate and 3 cookies on another they might answer 8. They used the count on strategy but included 6 when they attempted to keep track of the 3 cookies to be counted (6, 7, 8).

13.2 Realizing order does not matter

Your child might be able to count on, yet you notice they always start counting on from the first number they see or hear. For example, if you ask your child how many cookies there would be if you had 6 cookies on one plate and 3 cookies on another they would correctly answer 9.

When you asked them to explain how they did it they would probably say: 6, 7, 8, 9.

That is, they started with the 6 and counted on 3 more.

If you asked the reverse task: I had 3 cookies on one plate and 6 cookies on another plate they again would probably give the right answer. However it would probably take them longer to answer. If you asked how they worked it out you would find they had counted on but this time started with 3: 3, 4, 5, 6, 7, 8, 9.

Jamie can count things around the house. When his mother, Katelyn, gives him 6 dinner plates to place on the table and asks him how many there are he correctly counts "one, two, three, four, five, six".

When she gives him 3 more plates and asks how many there are in total he counts, "seven, eight, nine" and says: "Nine!". Katelyn knows this means Jamie can count-on, and she is pleased with his counting progress. But a few minutes later, something puzzled her.

She gave Jamie 3 knives to put out, and he counted "one, two, three".

Then she gave him 6 more knives, saying: "Here's 6 more."

To her surprise, Jamie counted on: "four, five, six, seven, eight, nine."

Katelyn was surprised because just a few minutes ago Jamie counted 6 and 3 more to get 9 plates.

Now he was counting 3 knives and 6 more to get 9. Why didn't he realize straight away that the answer was the same in both cases? Why didn't he see immediately that 3+6 is the same as 6+3?

Jamie, who is just starting to use counting-on, does not realize that 6 and 3 will give them the same answer as 3 and 6. This is because when he counts he is hearing different counting sequences: 7, 8, 9 compared to 4, 5, 6, 7, 8, 9.

At this stage in his counting development Katelyn needs to provide Jamie with tasks that require him to compare answers to tasks such as 6 + 3 and 3 + 6.

Jamie will be helped by starting with numbers that are close together in the sequence 1 through 10 - such as 3+4 and 4+3 - before giving him numbers like 8+2 and 2+8.

Once Jamie realizes that the order in which numbers are added does not affect the answer, he will count on from the larger number if he needs to use the count on strategy.

Chapter

Counting with units

Children make a very big conceptual step in their counting skills when they begin to count using units other than 1 - by two's, for example. Typically when children begin counting using units they use units of two, five, or ten.

14.1 Moving from counting-on

Children who can count-on - that is, begin counting from a given number - will usually use this strategy to figure out a problem such as 6 + ? = 10. They will also use counting-on to solve addition problems such as 6 + 5 = 11.

Once your child counts-on for most addition tasks you will want them to move to the next most efficient counting strategy. That is, you will want them to become a unit counter, and the first step in that growth stage is to count physical units.

If you have assessed that your child is proficient with counting-on and seems to be ready to become a physical unit counter you will want to provide activities that encourage them to count in groups rather than by ones.

14.1.1 Physical unit counting

A physical unit counter can count by groups - for example twos - and can work out how many times they have counted up to a given number.

Firstly, you can see how far they can verbally count by twos, fives and tens. You might start with tens as children find this counting sequence easier to remember: "10, 20,30,..."

To decide what your child already knows about counting by tens say to them: "Can you count by tens for me? Start at 10 and count as far as you can". You should record how far your child can count in their counting diary.

Next try the counting by fives. This has a repetitive pattern which is also easy to remember. To decide what your child already knows about counting by fives say to them: "Can you count by fives for me? Start at 5 and count as far as you can". Record how far your child can count in your diary.

Now try counting by twos. Say: "Can you count by twos for me? Start at 2 and count as far as you can." Again, record in your diary how far your child can count by twos.

Your child will benefit through practising counting by numbers other than one. This practice can include both oral and written activities.

You can ask your child to write down the counting sequences for you on paper to keep. You can glue or staple this into the diary. This way you can keep a check on how they progress over time.

Your child might enjoy writing the counting sequence on a piece of tape with a marker pen or crayon, or on a small whiteboard with a whiteboard marker.

The next step is to check how your child makes and counts groups of objects such as counting chips. You can use the following task to check what they do:

Hand your child a container with about 20 counters in it.

Ask your child to show you 2 groups of 5 counters.

Watch carefully what they do.

Does your child make 5 groups of 2 counters placed randomly on the table?

If your child makes 5 groups or rows they are not really doing as you asked. They have made 5 groups or rows of 2 counters and not the 2 groups of 5 objects as you asked. Although they both give a correct answer they mean different things and will look different.

After you have asked your child to make 2 groups of 5 objects observe how they find the total of the groups. For instance, does your child count by ones, count by twos or count by fives? If you aren't sure what they are doing ask them. You might like to record how they make, and count, the groups in your child's counting diary.

14.1.2 Unit counting in arrays & groups

Informally arranged groups of objects are not going to help your child move away from counting by ones. The examples below show objects arranged in *arrays*, that is rows and columns, or in equal-size *groups*, which structure the total number of objects so that physical counting by units is stimulated, and becomes a natural thing to do.

These are objects you might see around the house, or which you can find online. They are all useful for assessing if your child is ready to physically count by units.

Figure 14.1: Boots arranged in groups

How many boots do you see?

Counting by 2's: "2, 4, 6, 8, 10, 12, 14"

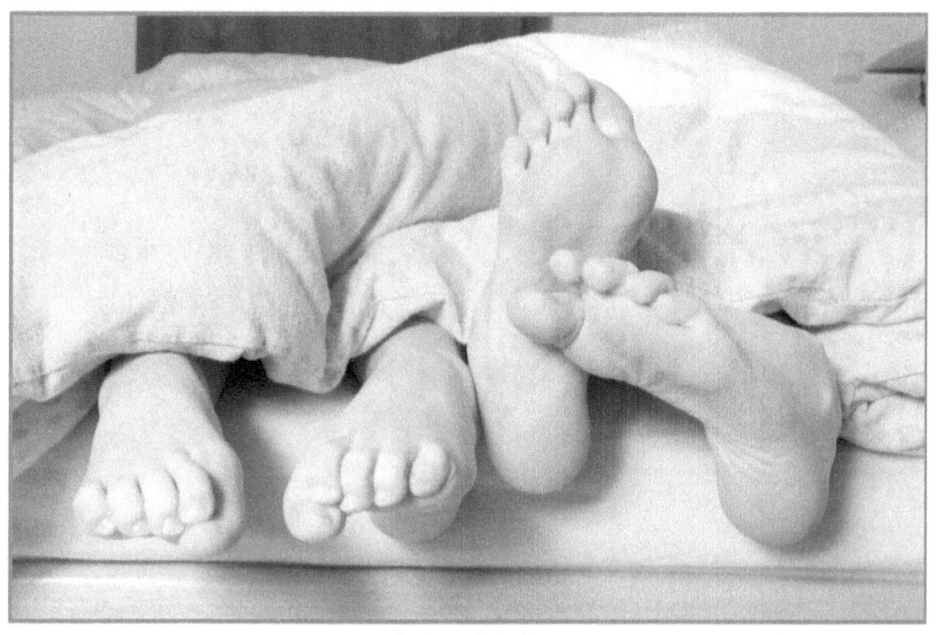

Figure 14.2: Toes in groups

How many toes do you see?

Counting by 5's: "5, 10, 15, 20"

Figure 14.3: Brick holes in an array

How many large holes do you see?

Counting by 2's: "2, 4, 6, 8, 10, 12, 14, 16, 18"

or by 3's: "3, 6, 9, 12, 15, 18"

or by 6's: "6, 12, 18"

Figure 14.4: Frog's toes in groups

How many toes do you see?

Counting by 3's: "3, 6, 9, 12"

Figure 14.5: Muffins in an array

How many muffins do you see?
Counting by 3's: "3, 6, 9, 12"
or by 4's: "4, 8, 12"

Figure 14.6: Emoticons in an array

How many emoticons do you see? Counting by 3's: "3, 6, 9, 12, 15" or by 5's: "5, 10, 15"

14.2 Children's understanding of arrays

As adults, we see a tray of muffins arranged into rows and columns:

Figure 14.7: A tray of muffins: 3 rows and 4 columns

We saw in the last section how useful it is to have objects arranged in arrays, or in equal-size groups (such as the frog toes), for children to physically count by units.

For the muffin tray, above we can count by 3s: "3, 6, 9, 12" or by 4s: "4, 8, 12".

But what leads us to think that young children actually "see" the rows and columns of muffins that we see?

What if children do not see the rows and columns? How then can they see how to structure the collection of muffins so as to count by units?

"Wait!" you say. "Surely any one, even young children, can see that the muffins are arranged into 3 rows each of 4 muffins?"

It turns out, through some beautiful studies by Lynne Outhred and Michael Mitchelmore, then at Macquarie University in Sydney, Australia, that many young children, around 5-6 years of age, *cannot* readily recognize rows and columns of an array.

You can read one of their articles at: https://bit.ly/2wbLZp7

Figure 14.8: Lynne Outhred & Michael Mitchelmore

One of the tasks given to children aged 6- 9 years of age was to cover a 12cm x 16cm rectangle (enclosed by a raised border) with 2cm x 2cm cardboard unit squares, and then *draw* the squares as an array. The researchers made the reasonable assumption, given he children's ages, that any difficulty in drawing the array would not be due to a lack of drawing skills, but due, rather, to a lack of understanding of the structure of an array.

What Lynne Outhred and Michael Mitchelmore found was that children displayed a wide range of abilities in reproducing arrays in drawings, with children sometimes doing better, or worse, as the tasks were changed a little.

Children would sometimes draw:

(*) an incorrect number of columns, with not all columns having an equal number of "squares":

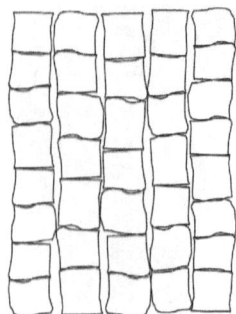

Figure 14.9: Incorrect numbers in columns, incorrect number of units

(*) a correct number of rows, with not each row having an equal number of "squares":

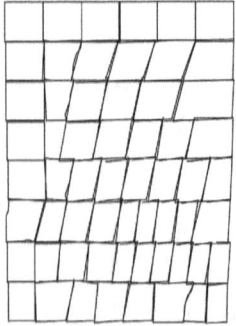

Figure 14.10: Correct number of rows, incorrect number of "squares" per row

You can see other examples of children's attempts at drawing arrays in the article at https://bit.ly/2wbLZp7

14.3 Separated arrays

The work of Lynne Outhred and Michael Mitchelmore throws into clear relief the problems young children have with placing tiles into rectangular arrays.

This gives us a lot of reason to pause and wonder whether children are seeing the rows and columns we see when we use arrays to count in units.

Figure 14.11: An array of hay bales

What, for example, might prompt a child to see the haystack, above, as a 6 × 6 array, and how would we know?

This a different skill to seeing, for example, 3 groups of 5 petals each in the following picture of flowers:

Figure 14.12: Three flowers, each with 5 petals

A child might mentally check, by counting from 1 through 5, that each flower has 5 petals, but then count by 5's to calculate how many petals there are in all: "5, 10, 15".

The only mental organization skill required for this is to "see" the petals in groups, which is rather easy since each group of 5 petals constitutes a "flower".

There is a difference between the arrays used by Outhred and Mitchelmore, on the one hand, and an array such as a tray of muffins, on the other. In the Outhred and Michelmore study the unit squares comprising the array were *contiguous*: the squares touched, or abutted, one another, as in the haystack example, above. In the muffin tray array, on the other hand, the muffins are separated to an extent that the spots they occupy are very clear: if we took the muffins out of the tray we would clearly see 12 empty holes, and the muffins themselves do not touch: they are separated one from another.

As you might suspect by now, we want to do two things:

1. *Assess* how, and to what extent, children can, or cannot, recognize and use arrays to count by units.

2. *Assist* children to recognize and use arrays to count by units.

14.3.1 Assessment

Generally, young children can see and identify identically-sized groups. For example, commonly young children will tell you that each of the frog's feet has 3 toes:

Figure 14.13: Three toes on each of the frog's feet

If children are not sure they can, and will, count the toes: "1, 2, 3".

Identification of equal groups - such as the number of toes on each frog foot - is less of an issue for young children than is identifying rows and columns of an array.

We can use children's ability to identify equal-size groups to transition into identification of equal-size rows or columns in a separated array.

For example, we can ask a child to show us the groups of 3 toes of the frog, and ask how many groups of 3 toes are there? Then we can show the muffin tray and ask: "Can you show me a group of 3 muffins?"

Figure 14.14: The tray of muffins: 3 rows and 4 columns

Watch carefully what your child does: do they show 3 muffins in a column, or do they simply show 3 muffins somewhere else - as the first 3 muffins in a row, for example.

If they show you 3 muffins in a column, ask them how many groups of 3 muffins they see.

If they did not show you 3 muffins in a column, show them the 3 muffins in a column and ask how many groups like that they can see.

Similarly, ask your child if they can see groups of 4 muffins, (this time, across the rows).

What you are looking for in this exercise is whether there is evidence that your child "sees" the columns and rows of the muffins as plainly as they see the groups of frog toes.

14.3.2 Assistance

Children can be helped to identify groups, on the one hand, and rows and columns of a separated array, on the other, through a structured micro-focusing on a group, a row, or a column.

What do we mean by this?

One way to focus on a *group* of frog toes, on one frog foot, is to cut out a small circle in a sheet of paper, or card, that shows just a group of 3 toes:

Figure 14.15: Focusing in on a single group of frog toes

After you do this you can ask your child to use the card to show you the other groups of 3, and again ask: "How many groups of 3 are there?"

At this point it is not necessary to ask what is the total number of toes, although you can, of course! The issue for now is whether your child is able to focus on the groups.

Asking them how many groups they see is a way of both focusing them on the groups and checking they are doing just that.

For a *separated array* you can cut out a rectangle in a piece of paper, or card, that shows just one row, or one column:

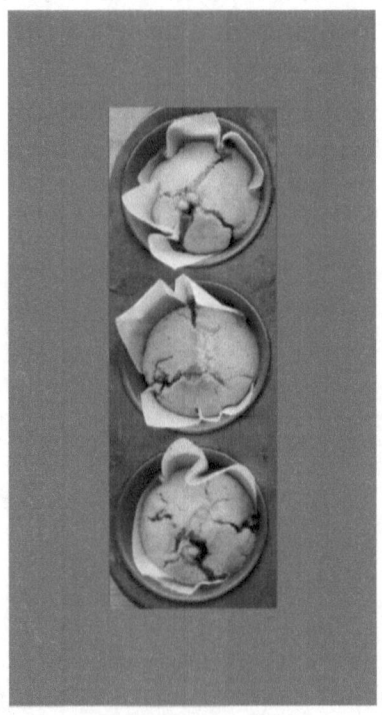

Figure 14.16: Focusing in on a single column of muffins

You will want to draw your child's attention to the column you have highlighted: if "column" is beyond their vocabulary the term "line" works well. Ask them how many lines like this they can see. Again, the focus for you, as parent or care-giver, is on your child's perception and identification of the columns of the array: asking them how many is a way of checking that they are, in fact, seeing the columns as such.

You can (and should!) repeat this with the rows of the separated array:

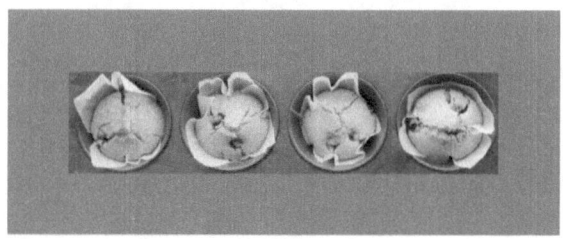

Figure 14.17: Focusing in on a single row of muffins

14.4 Figurative unit counting

Eva is 6 years old, and she can count by twos. She showed her aunt Penny how quickly she could count: "2, 4, 6, 8, 10, 12, 14, 16, 18, 20, 22, 24, 26, 28, 30, 32," She rattled off. "I can count higher if you want."

Penny laughed. "No, that's great. You can sure count pretty good."

"You know", said aunt Penny, "I was counting by twos the other day, and I counted to 24. How many times do you think I counted?"

"Wait!" said Eva, and she looked up to her right, tapping two fingers rhythmically as she nodded her head and blinked.

"12" she said to her aunt.

"12!" exclaimed Penny. "How did you figure that?"

"I just counted," said Eva.

But what did Eva count? Luckily we have a clue, because her aunt asked her:

"What did you count? I saw you tapping your fingers and looking up into the air."

"I was counting how many times I did this," said Eva, showing her aunt how she tapped her leg with two fingers.

Eva seemed to be counting how many times she counted by two:

Counting by 2s:	2	4	6	8	10	12	14	16	18	20	22	24
Times counted:	1	2	3	4	5	6	7	8	9	10	11	12

Figure 14.18: Number of acts of counting by 2s to 24

When she is counting, Eva focuses on her acts of counting by two.

In other words, Eva understands counting by two well enough to be able to take a single act of counting by two as a thing to be counted.

Physical counters need to see, touch, or hear physical objects in order to count them. Eva can focus on a physical action - her act of counting by two - and take that as a thing she can count.

This is a big step in Eva's mathematical development. Think about it - Eva is really doing division by 2: she can work out how many twos there are in a given even number.

This what Eva can presently do. She might already be able to count by tens and, with some more experience and a little help, count by fives. What we expect when Eva counts by ten or by five is that she will be able to count how many times she counts.

Eva is a figurative unit counter. She can count by groups - for example twos - and can work out how many times she has counted up to a given number, and she can do all that without counting physical objects. However, she cannot yet take the memory of doing that and use it to count on by twos from a given result.

Eva's mom asked her how many rows of 2 she could make with 10 counters. Her mom had exactly 10 counters ready to give to Eva if she asked, but not until she asked. In fact, Eva did not ask for the counters. She answered: "5 rows". Eva could work out, in her head, how many rows of 2 there would be without using the physical counters.

When she had made 5 rows, her mom asked: "If you had 28 counters how many more rows of 2 could you make?"

Eva could not figure out the answer to this question in her head. She asked her mom for some counters to work it out.

Eva is a figurative unit counter, and at this stage of her counting development she cannot answer this question. A figurative unit counter behaves just like a figurative counter with respect to counting groups - such as groups of two. A figurative unit counter always needs to begin counting from the first group of two: "1 group of two, 2 groups of two".

To answer the question that Eva's mom asked, a child must be able to count-on by groups of two rather than simply count on by ones. In a very real since, Eva counts by twos as a figurative counter would count by ones. She cannot yet count-on by twos.

14.5 Counting-on by units

Alysha is 8, and her dad is laying tiles in the bathroom. Alysha is helping bring in the tiles for her dad to lay, as he spreads adhesive on the floor. Her dad looks at the tiles already laid and says out loud: "I hope we will have enough tiles to finish the floor."

Alysha can see the tiles are in rows of five. Her dad has already laid 6 rows.

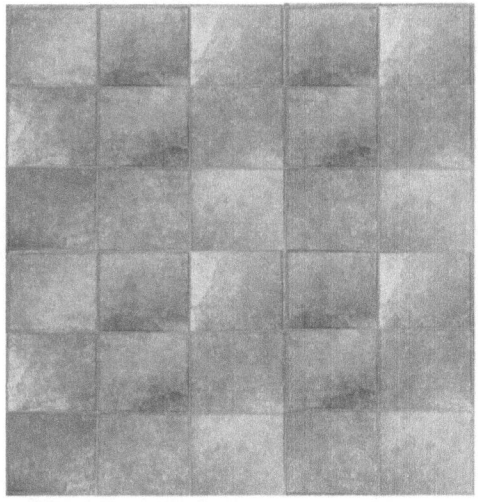

Figure 14.19: Alysha sees 6 rows, each of 5 tiles

"Yes, we do have enough tiles," she says.

"How did you figure that?" asks her dad. He knows Alysha is good at arithmetic but he is surprised how quickly she worked out they would have enough tiles.

"Because there's 6 rows on the floor, and, well ... that's 30 tiles ... but, see, we could make another 5 rows: 35, 40, 45, 50, 55, because we've got 55 tiles ... if you include the ones we already laid, I mean."

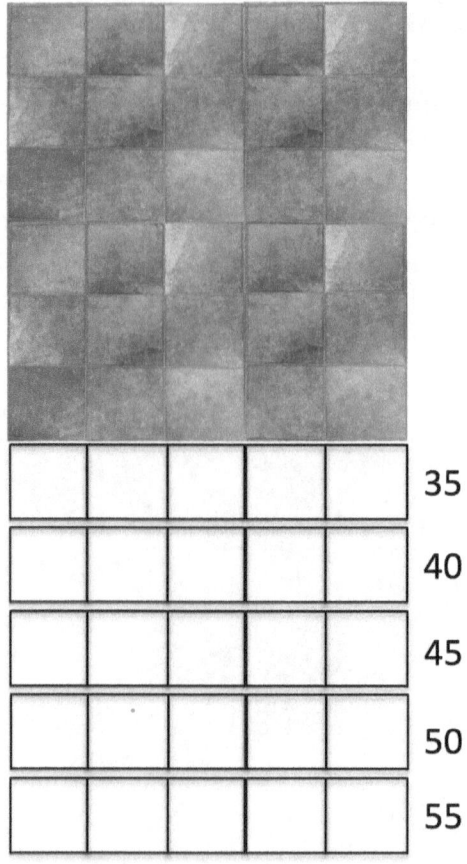

Figure 14.20: Alysha counts by 5's, to add 5 more rows

Alysha is able to count on by multiples of 5 from a given number. She can start from a number, count by fives, and keep track of how many times she has counted by fives. Unlike Eva in the previous section, Alysha does not have to begin her counting by fives from the number group [1, 2, 3, 4, 5]. She is able to start from the number group [31, 32, 33, 34, 35]. Alysha does not count "31, 32, 33, 34, 35". She is able to count by fives: so she simply counts "35, 40, 45, 50, 55". What Alysha can do that Eva cannot is start counting by fives from a number other than 5. Eva always has to begin from 5 - Alysha does not.

Alysha's counting abilities are impressive but there are still counting questions that are difficult for her. Alysha can count by fives from a number like 13: "13, 18, 23, 28, ..." but she finds it much harder than counting on from a multiple of 5.

Alysha also still finds it hard to answer a question like: "How many rows of tiles would we have if we had 24 tiles in all, and each row had 6 tiles?"

The problem for Alysha is that she is not good at counting by sixes yet, and it's not easy for her to count "6, 12, 18, 24" and at the same time keep track of how many times she has counted by sixes.

To progress in her counting development Alysha needs her counting-on-by-units skills extended and reinforced.

Chapter 15

Writing numbers

15.1 Numbers are everywhere

Typically we write numbers in the base 10 system using the digits 0, 1, 2, 3, 4, 5, 6, 7, 8, 9.

There are regularities in this system, which are not immediately obvious to young children, and remain hidden from many children for a long time. These regularities have to do with the fact that our system of representation of numbers is a *place value* system.

When we write 173, for example, we, as adults, have learned that the digit "1", because of its place, really stands for 1×100 and the digit "7". because of its place, really stands for 7 × 10, while the digit "3" because of its position (in the units place) simply stands for "3".

Long before they can count fluently, young children encounter written forms of numbers in many places: at a supermarket, on power poles, on road signs, in telephone numbers, on house addresses, and on birthday cards, for example.

Figure 15.1: House numbers

Children learn about numbers by learning to count. They also learn about them as artifacts in the adult world. In the early years of schooling teachers generally stick to counting up to small numbers. But children have another sense of number and magnitude and meaning that comes from the way numbers are used in the world outside of school.

Children are aware at a young age that houses have numbers, and they usually know the number of their house. By around age 4 - 5 you can ask them: " Would there be a house in your street numbered 2400?" (or some large number clearly beyond the numbers of houses in their street).

Their answer - a confident "no", or a thoughtful "not sure" - will tell you if they have a sense that 2400 is a large number compared to the house numbers in their street.

Figure 15.2: Supermarket numbers

Numbers are everywhere, and long before they can count confidently, children are trying to make sense of the number patterns they see in everyday life.

15.2 How children write numbers

Children around 4-5 years of age can hold apparently quite contradictory ideas about written forms of numbers.

For example, it is common for young children to tell you - an adult - that numbers in the hundreds have "three numbers" (meaning three digits).

This is a regularity in the patterns of written numbers that they have discovered, or been shown or told - most commonly, discovered for themselves.

Yet along with this nice observation young children when asked will often write a number such as "one hundred seven" ("one hundred *and* seven" for British English readers) as:

$$1007$$

Figure 15.3: Young children may write "one hundred seven" this way

They do this presumably because they know that "one hundred" is represented as "100" so they simply tack on a "7". This does not seem to cause any contradiction, or clash, with their stated position that numbers in the hundreds have "three numbers".

In our experience this is not something to "correct". Children who do this are finding ways to express their ideas about numbers, especially larger numbers, in written form. They are simply trying to find ways and means to make sense of, end express their thoughts on, larger numbers.

In the next section we discuss a game devised by Corina Silveira, then a doctoral student at the University of Southampton in the UK, as a simple fun game for 3 or more players that gives an adult observer knowledge about how children see the representation of numbers.

You can extend this game, as we indicate at the end of the next section, to assess and address children's familiarity with the written form of numbers in the hundreds.

15.3 Which is the biggest number?

To play this game you will need several children - four is a good number - about 4 - 5 years of age. Several parents could get together and have their children play this game.

You will also need a pack of blank cards on which you have written the numbers 1 through 100 (one number per card):

Figure 15.4: Numbered cards

One of the children deals a small number of cards - 4, for example - to the others, and themself, and the children take turns in placing a single card , face up, on the table. The child who placed the largest number wins the hand.

The question you, as adult, are interested in is: how do the children decide which is the largest number?

An example from Corina's own work was when she asked a group of children how they knew 63, on a card, was bigger than 29, as they said it was.

One child pointed to the card with 63 and said:
"Because it's got a 6."
When Corina objected that the other card had a 9, the children jostled to tell her, with almost pity in their eyes that an adult would not know such a thing:
"But the left hand's the boss!".

These children could not read 69, but knew that the left hand 6 meant it was a bigger number than 29.

As you play this game with children you will often see things that give you insight into how children recognize written numbers. For example, when Corina Silveira was first playing this game with children, a few times the card with "100" on it was dealt, and each time that happened the child who got the card started laughing. One boy laughed so hard he fell backwards onto the floor, crying out:"Oh, oh! It's the biggest one of all."

These children are showing a beginning awareness of the role of the place of the digits in determining the size of a number.

This is an assessment task, so please try to approach it that way. It is not a task for you to "teach" children reasons that adults know which numbers are bigger Rather, it is an assessment task for you to discover exactly how it is that children themselves are reasoning about the size of numbers. It is your knowledge of their procedures, and their thoughts, that will enable you to help them later.

Remember: compassionate assessment through observation first; help through mastery, later.

15.3.1 Numbers in the hundreds

For a more advanced form of this game, suitable for children with are already discussing and thinking about numbers in the hundreds is to make a pack of cards with the numbers 100 though 300:

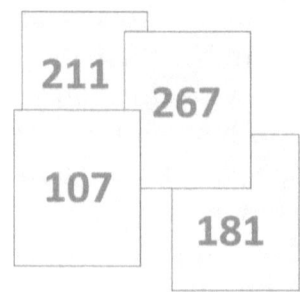

Figure 15.5: Cards with numbers from 100 through 300

Again, as children play this game you are watching to see just how it is that they make sense of the form of larger numbers, and how they decide which of a group of such numbers is the largest.

It is their thought patterns you are trying to discover - the ways they are making sense of larger numbers.

And you are doing this not to "correct" them if they do not do it the adult way, but to give you insight into how they are thinking so that you may devise new and interesting number problems for them, to advance their thinking and skills.

The trick is to avoid telling them how they should behave as little adults, and instead to enter the world of their thoughts - as strange as those thoughts may seem to you at times! - because it is by understanding where they are coming from, and showing them you understand, that prompts them to pay attention to you.

15.4 How many letters are there?

Here is a cute game for children to explore that connects numbers to their written names:

1. Pick a number, for example: "27".

2. Write the number in the child's language (in our case, English): "twenty seven"

3. Count the number of letters in the number name: "11".

4. Repeat until you keep getting the same number.

The numbers, and their (English) names you get are:

27	twenty seven
11	eleven
6	six
3	three
5	five
4	four
4	four

What if we started with another number, say "eighty nine"?

This has 10 letters and the (English) name for "10" is spelled "ten", which has just 3 letters.

But we saw already in the table above, that 3 leads to 5 and then to 4.

Try it with larger numbers, such as 163. Remember that the spelling of "163" in American English is "one hundred sixty three" whereas in British English it is "one hundred and sixty three". Does it make a difference? How?

What if we did the first example in German instead of English?

27	siebenundzwanzig
16	sechszehn
9	neun
4	vier
4	vier

What happens in other languages? French, Korean?

If you don't speak or read German, French or Korean you can use Google Translate to help you out.

This fun game combines simple counting by 1s with a numeration system - a way of writing numbers using the digits 0, 1, ..., 9 and place value - together with the ways of writing numbers using language.

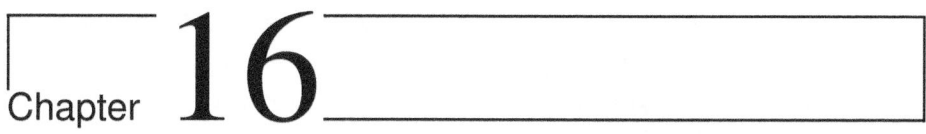

Chapter 16

Addition and subtraction

As a parent you want to see your child be able to carry out addition and subtraction problems with ease, as they progress through elementary school.

The good news is that the issues with which we have been dealing - counting-on and counting by units - are exactly what a child needs to master in order to develop facility with addition and subtraction!

Figure 16.1: Preparing for addition

That's right: we have lead you to this point so that your child will have a head start in the basic operations of arithmetic.

Skill with counting, and with the more advanced forms of counting we have detailed, is critical for your child to develop ease with addition and subtraction, typically in 3rd and 4th grades.

Let's see, in some detail, just how counting skills factor into skill with addition and subtraction.

We have seen how a child who can count-on can solve the "How many?" question:

$$6 + ? = 13$$

by counting-on from 6 to 13, while keeping track of the seven counts using fingers, for example. Sometimes children will answer this question using very slight body movements - such as biting their lip - instead of using fingers.

A child who is able to count-on by units, by contrast, might see that 6 and 6 more is 12, so the answer is 1 more than 6. Thinking like this requires a very flexible use of units:

- The child sees the "6" as a unit able to be repeated to get a known result of "12".

- The child sees the "13" as a unit that can be split into a unit of "12" and a unit of "1".

- the unit of "1" can be combined with the unit of "6" to give a unit of "7".

Of course there are simpler questions like this that are within the range of children who are just learning to count with units:

$$2 + ? = 8$$
$$5 + ? = 20$$

Janine is quite good at counting by 2's. She knows that to count to 24 by 2's would take 12 steps. So she can answer the how many question:

$$2 + ? = 8$$

by recognizing that it takes 4 steps to count to 8 by 2's, and so 3 steps to count from 2 to 8, giving a result of 3 lots of 2, or 6.

Rory answered the "How many?" question:

$$5 + ? = 20$$

by saying out loud: "Five, ten, fifteen, twenty" then pausing for a moment to say: "Fifteen!"

When his dad asked him how he figured that Rory said: "Well, you see, it's like five, ten, fifteen, twenty - that's four, so from the five that's three. So three fives, that's fifteen."

We infer that Rory is able to flexibly count by 5's, to recognize that he would need three extra steps to go from 5 to 20, and to know, or to figure by figurative unit counting, that 3 lots of 5 is 15.

16.1 Dominoes and number facts

You can help your child build a store of remembered number facts through playing games with dominoes.

Figure 16.2: Dominoes

16.1.1 Simple partitions of numbers

Here's one simple game to play:
How many different dominoes are there whose total number of dots add to 8? Remember that dominoes can have 0 dots on one or both sides.

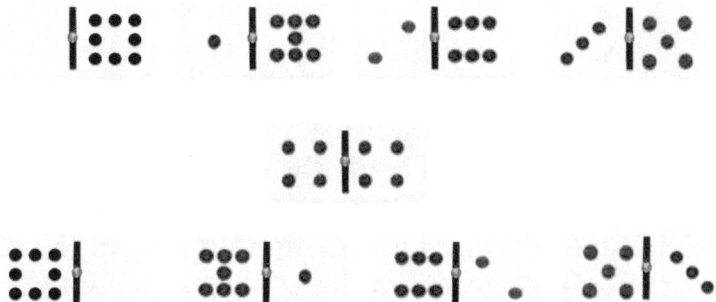

Figure 16.3: All dominoes whose dots add to 8

Each domino in the top row of 4, when flipped horizontally, gives a domino in the bottom row, and vice versa, so there are only 5 different dominoes whose dots add to 8, corresponding to the number facts:

0 + 8 = 8 = 8 + 0
1 + 7 = 8 = 7 + 1
2 + 6 = 8 = 6 + 2
3 + 5 = 8 = 5 + 3
4 + 4 = 8

You can play this number structuring game with your child, using numbers within their counting or recognition range.

Encourage them to write the addition facts as above.

16.2 Partitions

A wonderful game to play with children is to ask them to figure all the different ways of writing a number as a sum of whole numbers.

For example, the number 4 can be written a a sum of whole numbers as follows:

4
3 + 1

2 + 2
2 + 1 + 1
1 + 1 + 1 + 1

Of course we could also write 4 as 1 + 3, but, as your child will probably tell you, that's really the same as 3 + 1.

These different ways of writing a number are called the partitions of the number, so we are looking to figure all the different ways of partitioning a given number.

A lovely way to visualize the partitions of a number is to use dots, or counting chips, arranged in rows. For example for the five partitions of 4, above, we can represent these different partitions using counting chips as follows:

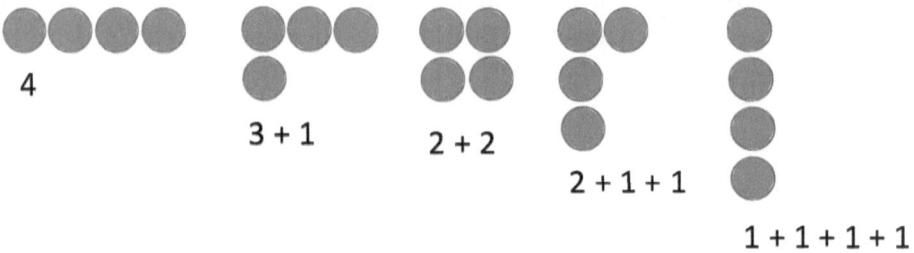

Figure 16.4: Partitions of 4

Playing with partitions, and trying to find all partitions of small numbers - up to 10, for example, helps your child focus on and recall basic addition facts.

Number facts, including addition facts, are best learned, and recalled, as part of fun games, involving exploration.

Recalled number facts are key to your child's fluency with arithmetic. For example, a child who can recall that 8+2=10 is very likely to be able to see straight away that 18 + 12 = 30, by seeing that 18+12 = 10 + 8 + 10 +2.

Addition facts are like stepping stones that, once recalled, can be used in other addition problems. For example a child who recalls that 9+1= 10 is likely to see, without further counting or calculation, that 9+8 = 9+1+7 = 17.

Recalled addition facts ease a child's working memory in addition problems.

Remember Annette Karmiloff-Smith's work that shows children are more likely to show growth and development under conditions of success when they are skilled in carrying out actions in a prior stage of their development.

What this means for addition problems is that children are more likely to be successful if they have behavioral mastery of simple addition facts. The more simple addition facts they can bring to mind easily, the more they free their working memory to see new number facts and connections.

16.3 Exploding dots

Numbers have to be written in some form, and we typically use the base 10 numeration system to represent numbers.

This is a place value system in which the digits 0, 1, 2, 3, 4, 5, 6, 7, 8, 9 are used in some order to represent multiples of 1, 10, 100, 1000, and so on.

For example, 7428 represents $7 \times 1000 + 4 \times 100 + 2 \times 10 + 8$ in a base 10 place value system

James Tanton uses a model for representing numbers in our base 10 system that he calls "exploding dots".

Figure 16.5: James Tanton

The basic setup is to think of "dots" - which can be physical counters- being placed in a series of boxes that extends arbitrarily far to the left:

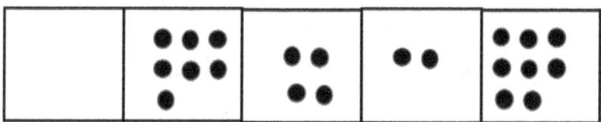

Figure 16.6: Exploding dots: first step

As you might imagine the numbers of dots in the boxes are there to represent the base 10 number 7428, but we do not have to explicitly say that "7", for example, represents 7×1000 because, as you will see, it is built-in to the "explosion" rule for this "machine".

We follow James Tanton in building a $1 \leftarrow 10$ exploding dots machine, in which there is one simple rule: every time there is a bundle of 10 dots in a box, the bundle "explodes" to a single dot to the box to the left:

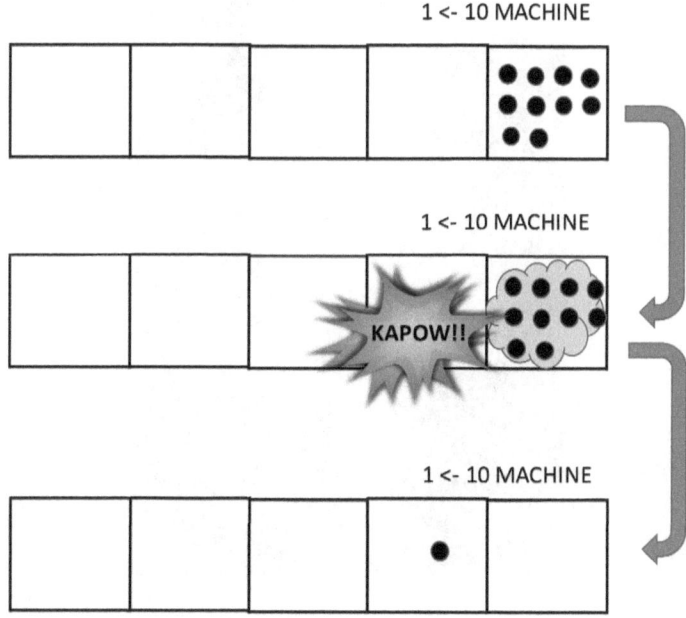

Figure 16.7: Exploding dots: $1 \leftarrow 10$ machine

That single dot in the second box from the right came about by an "explosion" from a bundle of 10 dots in the box to the right.

16.3.1 Base 10 exploding dots : place value

A wonderful feature of a $1 \leftarrow 10$ exploding dots machine is that it gives an action-based physical meaning to place value in our base 10 way of writing numbers.

Many children struggle in school arithmetic to understand, in subtraction exercises such as 403-137, that the "4" represents 4×100.

Here is a useful activity for you to assess your child's understanding of place value. It is based on a similar interview question of Marilyn Burns:

Figure 16.8: Marilyn Burns

> From a large pile of counting chips ask your child to count out 27 chips.
>
> Then ask your child to write the number "twenty seven" for you.
>
> Now ask your child to show you, using the chips, what the "2" means when they wrote "27".

If your child answers immediately that it means "twenty" you can feel confident that they have a good beginning appreciation of place value.

Your child may count out twenty of the chips and show you, or perhaps then say "twenty". This also indicates a beginning appreciation of place value, though not expressed as confidently.

If you child indicates anything other than "twenty" this gives you an indication that they are not yet comfortable in their understanding of place value.

Playing with numbers in a 1 ← 10 exploding dots machine makes this perfectly clear, not by way of "explanation", but simply through experience. And like all successful learning through experience, there is an important role for a parent or care-giver in helping a child focus attention and form long-lasting relevant memories from experience.

Exploding dots is most useful when your child can already count reliably into the 20's, 30s, 40s and beyond. The reason is simple: "explosions" in a 1 ← 10 exploding dots machine do not happen until a box has 10 or more dots.

A simple way to start exploring the base 10 place value system is to play with simple addition problems using exploding dots. Here are some examples you can try:

"Here's 9 farm animals, and in the box on the table there are 8 more? Can you show me, using exploding dots, how many there are altogether? How we would we write that?"

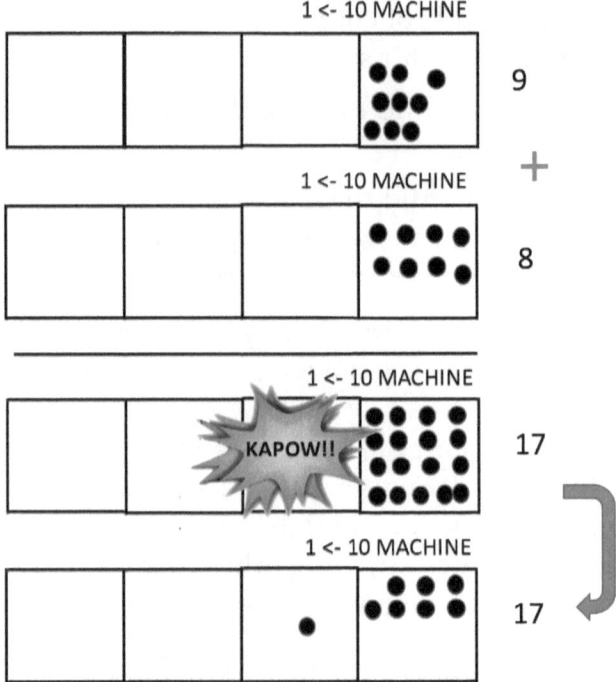

Figure 16.9: 9+8 in a 1 ← 10 machine

Your child may already be able to count-on, in which case they may simply answer "17" and explain to you that they counted: "ten, eleven, twelve, thirteen, fourteen, fifteen, sixteen, seventeen". Even so, asking them to show you how to calculate 9+8 using a 1 ← 10 exploding dots machine focuses their attention on *why* we write "seventeen" as "17" and that the "1" in "17" really stands for 1×10.

You can try other examples suited to your child's ability to count. If they are fluent with larger numbers try them with 19+12, 25+17, for example, using a 1 ← 10 exploding dots machine.

The aim of this activity is not so much to figure out the result of an addition, as it is to focus your child's attention on how, and why, we write the answer (in base 10) the way we do.

Remember, understanding place value is not a trivial matter. The great Greek mathematician Archimedes was not familiar with place value, and many children suffer in arithmetic problems in school because they have no understanding of place value and the base 10 notation system. Exploding dots can fix that in a super fun way!

16.3.2 Exploding dots: Addition

We can use a 1 ← 10 exploding dots machine to help us do addition, that is normally out of reach for younger children, who are still learning to count.

Let's see how we might use an exploding dots 1 ← 10 machine to figure out which base 10 number is 43 + 29:

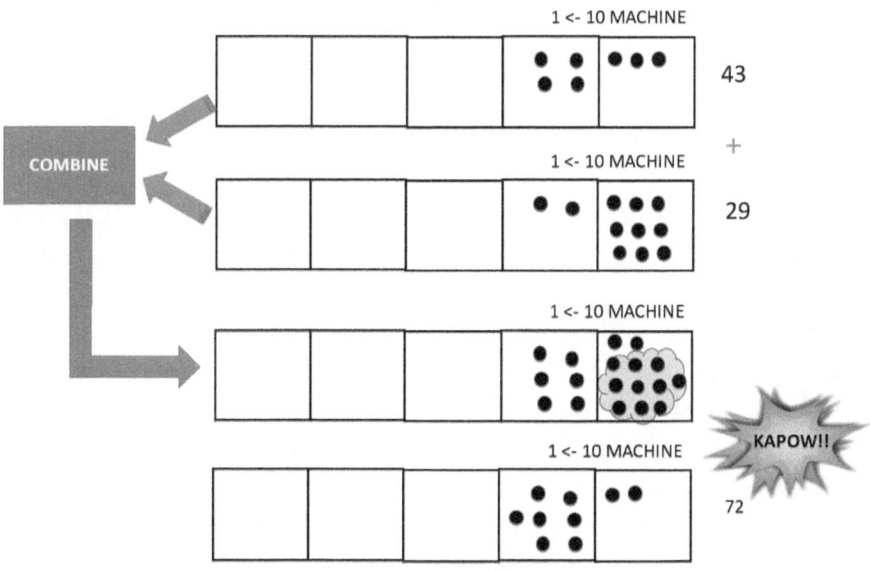

Figure 16.10: Exploding dots: 43+29 = 72

How does a child know that 43 is represented in a 1 ← 10 exploding dots machine as 3 dots in the right-most box and 4 dots in the box to the left?

Initially, they may not. But, so long as they can count to 43, they can place 43 dots in the right-most box and do four "explosions":

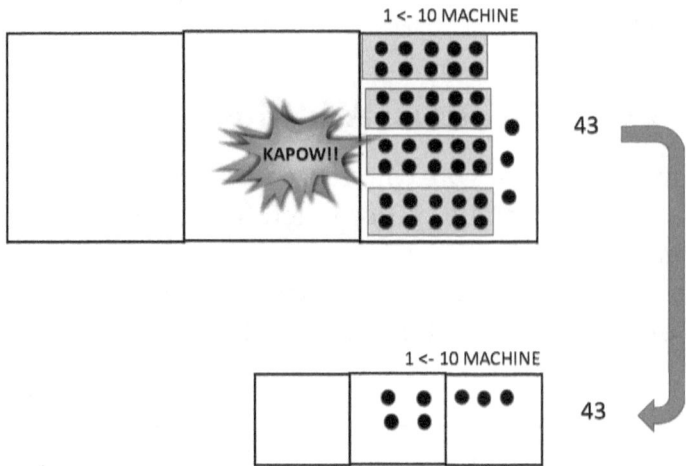

Figure 16.11: Exploding dots: Exploding 43 dots

These exploding dots activities can be done successfully with children to the extent that you use numbers to which they can count.

For example, in the case of 49+23, above, if a child is not yet understanding place value - which early on, most will not - then they need, in that example, to be able to count to 43. If they cannot yet count that far then limit the additions to numbers to which they can count.

Exploding dots will help with both addition skills and a deep understanding of, and facility with, place value.

Once a child has an appreciation for place value - essentially, an awareness that a dot in a box (except the units box) can be "unexploded" to 10 dots to the right - then even much more complicated additions become relatively easy:

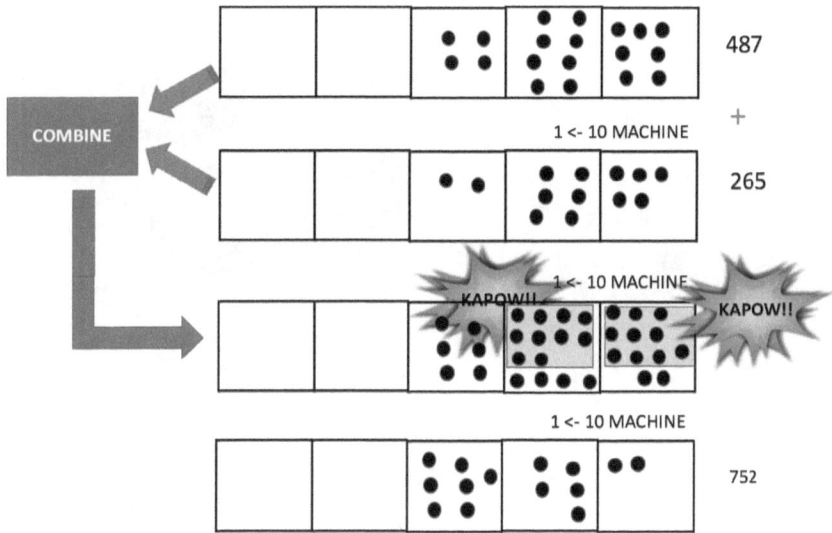

Figure 16.12: Exploding dots: 487 + 265

16.3.3 Exploding dots: Subtraction

Sometimes subtraction is called "take away" by elementary teachers and students.

Exploding dots makes it pretty clear why this name is appropriate, and makes subtraction problems ("take aways") fairly easy, even when "borrowing" is involved.

Let's think for a moment what a subtraction problem like

$$8 - 5$$

is asking.

It is asking for a number, which we don't know yet, but will temporarily name with a question mark "?", such that

$$? + 5 = 8$$

If we had 8 counters, for example, and took away 5 of them we would have 3 left:

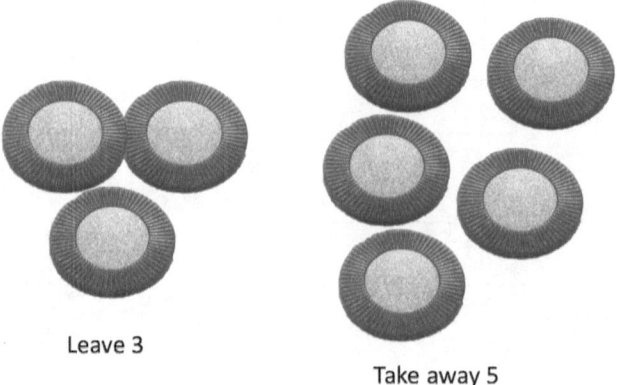

Figure 16.13: 8 - 5 = 3

This would play out the same in a 1 ← 10 exploding dots machine:

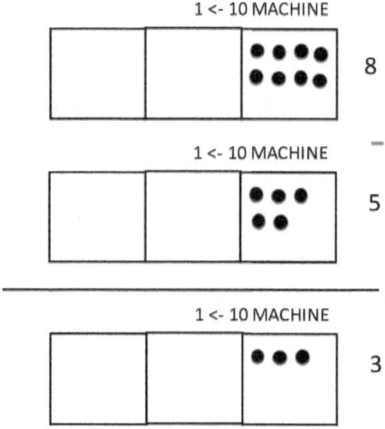

Figure 16.14: Exploding dots: 8 - 5 = 3

A major feature of a 1 ← 10 exploding dots machine is that it allows children to do subtractions that are apparently much more difficult.

For example, suppose we want to subtract 27 from 62:

62 - 27

We can set this up in a $1 \leftarrow 10$ exploding dots machine in the following way:

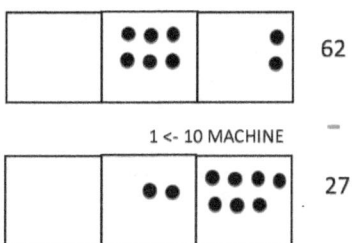

Figure 16.15: Exploding dots: 62 - 27

What if your child can count to 62 but does not yet quite grasp the base 10 numeration system? In that case ask your child to put 62 dots in the right-most box and explode any groups of 10 into the box to the left. Do the same thing with 27.

When your child can represent 62 and 27 in a $1 \leftarrow 10$ exploding dots machine, as above, there's a couple of different paths they can take.

First, you and your child can see that we cannot take the 7 in the units box for 27 from the 2 in the units box for 62. But we can fix this by "unexploding" one of the 6 dots in the 10s box for 62 to 10 dots in the units box:

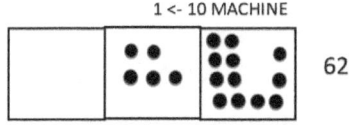

Figure 16.16: Exploding dots: 62 unexploded to fifty-twelve!

This is just like saying we can think of "sixty-two" as "fifty-twelve". And now we can subtract:

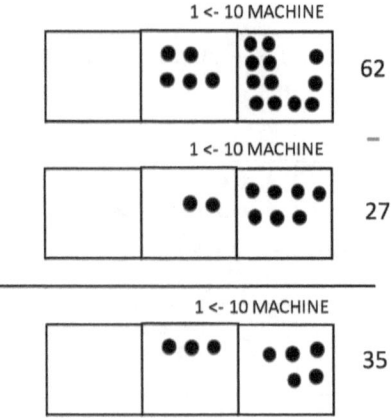

Figure 16.17: Exploding dots: 62 -27 = 35

Or, you and your child could have subtracted the 2 dots in the 10s place for 27 from the 6 dots in the 10s place for 62, and then unexploded one of the remaining 4 dots in the 10s place for 62 to 10 dots in the units place for 62.

Depending on your child's level of comfort and fluency with the base 10 numeration system - what the dots in the units, 10s, 100s, ... boxes represent - even apparently more complicated subtractions are relatively easy:

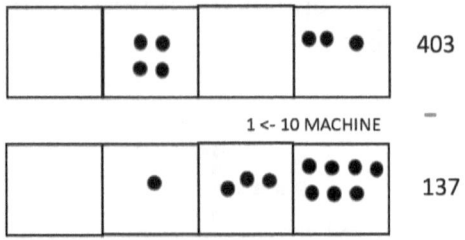

Figure 16.18: Exploding dots: 403-137

You can see the 3 dots in the 10s place for 137 is going to cause a problem in subtracting from 403, where there are *no* dots in the 10s place.Unless ... we unexplode one of the dots in the 100s place for 403:

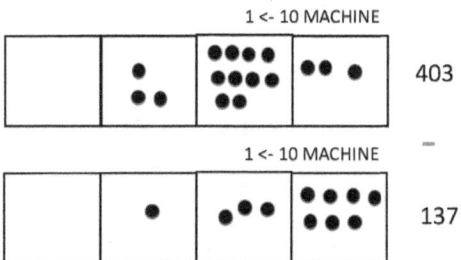

Figure 16.19: Exploding dots: 403-137, with 403 written as "three hundred and tenty three"!

Now it's the units places that need some adjustment to allow us to easily subtract, so we unexplode one of the dots in the 10s place for 403 to 10 dots in the units place, and then we can finish our take away:

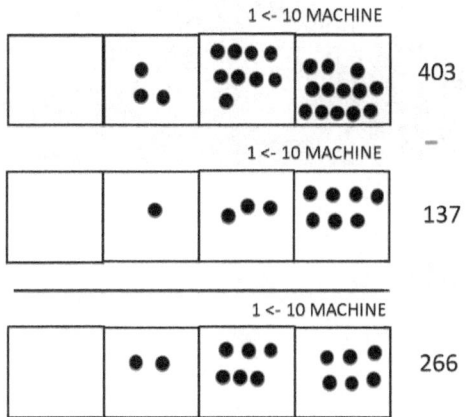

Figure 16.20: Exploding dots: 403-137= 266

16.3.4 Wait ... there's more!

So powerful is the idea of a $1 \leftarrow 10$ exploding dots machine that the whole of school arithmetic can be based on it. Exploding dots machines make multiplication, division, and even algebra, problems as simple, and fun, as ordinary addition and subtraction.

You can see and read much more at the following places:

1. James Tanton's own website:

   http://gdaymath.com/courses/exploding-dots

2. The Global Math Project: https://www.explodingdots.org

www.ingramcontent.com/pod-product-compliance
Lightning Source LLC
Chambersburg PA
CBHW020236170426
43202CB00008B/110